How to Solve Word Problems in Algebra

A Solved Problem Approach

Second Edition

Mildred Johnson
Late Professor Emeritus of Mathematics
Chaffey Community College
Alta Loma, California

Timothy Johnson
Linus Johnson
Dean McRaine
Sheralyn Johnson

McGraw-Hill

New York   Chicago   San Francisco   Lisbon   London
Madrid   Mexico City   Milan   New Delhi   San Juan
Seoul   Singapore   Sydney   Toronto

Library of Congress Cataloging Card Number: 99-31719

McGraw-Hill

A Division of The McGraw·Hill Companies

17 18 19 20 21 22 23 24 25   DOC/DOC   1 5 4 3 2 1

ISBN 0-07-134307-5

*The sponsoring editor for this book was Barbara Gilson, the editing supervisor was
Maureen B. Walker, and the production supervisor was Elizabeth J. Strange. It was set in
Stone Serif by PRD Group.*

Printed and bound by R. R. Donnelley & Sons Company.

McGraw-Hill books are available at special quantity discounts to use as premiums
and sales promotions, or for use in corporate training sessions. For more information, please write to the Director of Special Sales, McGraw-Hill, 11 West 19th
Street, New York, NY 10011. Or contact your local bookstore.

This book is printed on recycled, acid-free paper containing
a minimum of 50% recycled, de-inked fiber.

Contents

Preface

There is no area in algebra which causes students as much difficulty as word problems. Most textbooks in algebra do not have adequate explanations and examples for the student who is having trouble with them. This book's purpose is to give the student detailed instructions in procedures and many completely worked examples to follow. All major types of word problems usually found in algebra texts are here. Emphasis is on the mechanics of word-problem solving because it has been my experience that students having difficulty can learn basic procedures even if they are unable to reason out a problem.

This book may be used independently or in conjunction with a text to improve skills in solving word problems. The problems are suitable for either elementary- or intermediate-level algebra students. A supplementary miscellaneous problem set with answers only is at the end of the book, for drill or testing purposes.

MILDRED JOHNSON

How to Work Word Problems

If you are having trouble with word problems, this book is for you! In it you will find many examples of the basic types of word problems completely worked out for you. Learning to work problems is like learning to play the piano. First you are shown how. Then you must practice and practice and practice. Just *reading* this book will not help unless you *work* the problems. The more you work, the more confident you will become. After you have worked many problems with the solutions there to guide you, you will find miscellaneous problems at the end of the book for extra practice.

You will find certain basic types of word problems in almost every algebra book. You can't go out and use them in daily life, or in electronics, or in nursing. But they teach you *basic procedures* which you will be able to use elsewhere. This book will show you step by step what to do in each type of problem. Let's learn how it's done!

How do you start to work a word problem?

1. *Read the problem* all the way through quickly to see what kind of word problem it is and what it is about.
2. Look for a question at the end of the problem. This is often a good way to find what you are solving for. Sometimes two or three things need to be found.
3. Start every problem with "Let x = something." (We generally use x for the unknown.) You let x equal what you are trying to find. What you are trying to find is usually stated in the question at the end of the problem. This is

called the *unknown*. You must show and label what x stands for in your problem, or your equation has no meaning. You'll note in each solved problem in this book that x is always labeled with the unit of measure called for in the problem (inches, miles per hour, pounds, etc.). That's why we don't bother repeating the units label for the answer line.

4. If you have to find more than one quantity or unknown, try to determine the *smallest* unknown. This unknown is often the one to let x equal.
5. Go back and read the problem over again. This time read it one piece at a time. Simple problems generally have two statements. One statement helps you set up the unknowns, and the other gives you equation information. Translate the problem from words to symbols one piece at a time.

Here are some examples of statements translated into algebraic language, using x as the unknown. From time to time refer to these examples to refresh your memory as you work the problems in this book.

Statement	Algebra
1. Twice as much as the unknown	$2x$
2. Two less than the unknown	$x - 2$
3. Five more than the unknown	$x + 5$
4. Three more than twice the unknown	$2x + 3$
5. A number decreased by 7	$x - 7$
6. Ten decreased by the unknown	$10 - x$
7. Sheri's age (x) 4 years from now	$x + 4$
8. Dan's age (x) 10 years ago	$x - 10$
9. Number of cents in x quarters	$25x$
10. Number of cents in $2x$ dimes	$10(2x)$
11. Number of cents in $x + 5$ nickels	$5(x + 5)$
12. Separate 17 into two parts	x and $17 - x$
13. Distance traveled in x hours at 50 mph	$50x$
14. Two consecutive integers	x and $x + 1$

15. Two consecutive even integers x and $x + 2$
16. Two consecutive odd integers x and $x + 2$
17. Interest on x dollars for 1 year $0.05x$
    at 5 percent
18. $20,000 separated into two x and $20,000 - x$
    investments
19. Distance traveled in 3 hours $3x$
    at x mph
20. Distance traveled in 40 minutes $2x/3$
    at x mph (40 minutes = $\frac{2}{3}$ hour)
21. Sum of a number and 20 $x + 20$
22. Product of a number and 3 $3x$
23. Quotient of a number and 8 $x/8$
24. Four times as much $4x$
25. Three is four more than a number $3 = x + 4$

FACTS TO REMEMBER

1. "Times as much" means multiply.
2. "More than" means add.
3. "Decreased by" means subtract.
4. "Increased by" means add.
5. "Separate 28 into two parts" means find two numbers whose sum is 28.
6. "Percent of" means multiply.
7. "Is, was, will be" become the equals sign (=) in algebra.
8. If 7 exceeds 2 by 5, then $7 - 2 = 5$. "Exceeds" becomes a minus sign ($-$), and "by" becomes an equals sign (=).
9. No unit labels such as feet, degrees, and dollars are used in equations. In this book we have left these labels off the answers as well. Just refer to the "Let $x =$" statement to find the unit label for the answer.

EXAMPLES OF HOW TO START A PROBLEM

1. One number is two times another.
    Let x = smaller number
           $2x$ = larger number

2. A man is 3 years older than twice his son's age.
   Let $\quad x =$ son's age
   $\qquad 2x + 3 =$ man's age

3. Represent two numbers whose sum is 72.
   Let $\quad x =$ one number
   $\qquad 72 - x =$ the other number

4. A man invested $10,000, part at 5 percent and part at 7 percent. Represent interest (income).
   Let $\qquad x \ = \$ amount invested at 5 percent
   $\qquad \$10,000 - x \ = \$ amount invested at 7 percent

   Then

   $\qquad 0.05\ x \ = \$ interest on first investment
   $\qquad 0.07(10,000 - x) = \$ interest on second investment

   When money is invested, the rate of interest times the principal equals the amount of interest per year.

5. A mixture contains 5% sulphuric acid. Represent the amount of acid (in quarts).
   Let $\quad x =$ number of quarts in mixture
   $\qquad 0.05x =$ number of quarts of sulphuric acid

6. A woman drove for 5 hours at a uniform rate per hour. Represent the distance traveled.
   Let $\quad x =$ rate in miles per hour
   $\qquad 5x =$ number of miles traveled

7. A girl had two more dimes than nickels. Represent how much money she had in cents.
   Let $\qquad x =$ number of nickels
   $\qquad x + 2 =$ number of dimes
   $\qquad 5x =$ number of *cents* in nickels (5 cents in *each* nickel)
   $\qquad 10(x + 2) =$ number of cents in dimes (10 cents in each dime)

FACTS TO REMEMBER ABOUT SOLVING AN EQUATION

(These are facts you should already have learned about procedures in problem solutions.)

1. Remove parentheses first.
   Subtraction $\qquad\qquad\qquad\qquad -(3x + 2) = -3x - 2$
   Multiplication (distributive law) $\qquad 3(x + 2) = 3x + 6$
2. Remove fractions by multiplying by the lowest common denominator (LCD).

$$\frac{3}{x} + 4 = \frac{5}{2x} + 3$$

The LCD is $2x$. Multiplying both sides of the equation by $2x$,

$$6 + 8x = 5 + 6x$$

3. Decimals should be removed from an equation before solving. Multiply by a power of 10 large enough to make all decimal numbers whole numbers. If you multiply by 10, you move the decimal point in all terms *one* place to the right. If you multiply by 100, you move the decimal point in all terms *two* places to the right.

Example:

$$0.03\, x + 201.2 - x = 85$$

Multiply both sides of the equation by 100.

$$3x + 20{,}120 - 100x = 8500$$

Now let's try some word problems. In this book they are grouped into general types. That way you can look up each kind of problem for basic steps if you are using an algebra text.

Chapter 1

Numbers

Number problems are problems about the relationships among numbers. The unknown is a whole number, not a fraction or a mixed number. It is almost always a positive number. In some problems the numbers are referred to as *integers*, which you may remember are positive numbers, negative numbers, or zero.

EXAMPLE 1

There are two numbers whose sum is 72. One number is twice the other. What are the numbers?

Steps

1. Read the problem. It is about numbers.
2. The question at the end asks, "What are the numbers?" So we start out with "let x = smaller number." Be sure you always start with x (that is, $1x$). Never start off with "let $2x$ = something," because it doesn't have any meaning unless you know what x stands for. Always label x as carefully as you can. In this problem, that means just the label "smaller number," since no units of measure are used in this problem.
3. Read the problem again, one piece at a time. The first line says you have two numbers. So far, you have one number represented by x. You know you have to represent two unknowns because the problem asks you to find two numbers. So read on. Next the problem states that the sum of the two numbers equals 72. Most of the time it is good to save a sum for the equation statement if you can. A sum may also be used to represent the second unknown, as you

1

shall see later. The next statement says "one number is twice the other." Here is a fact for the second unknown!

Now you have

Let x = smaller number
$2x$ = larger number

4. Now that both unknowns are represented, we can set up the equation with the fact which has not been used. "The sum is 72" has not been used; translating it, we get the following equation:

$$x + 2x = 72$$

Now let's put it all together and solve.

Solution

Let x = smaller number
$2x$ = larger number

Then

$$x + 2x = 72$$
$$3x = 72$$
$$\left.\begin{array}{r} x = 24 \\ 2x = 48 \end{array}\right\} \quad \textit{Answers}$$

Check: The sum of the numbers is 72. Thus 24 + 48 = 72. *Be sure you have answered the question completely*; that is, be sure you have solved for the unknown or unknowns asked for in the problem.

EXAMPLE 2

There are two numbers whose sum is 50. Three times the first is 5 more than twice the second. What are the numbers?

Steps

1. First carefully read the problem.
2. Reread the question at the end of the problem: "What are the numbers?"

2

3. Note the individual facts about the numbers:
   (*a*) There are two numbers.
   (*b*) Their sum or total is 50.
   (*c*) Three times the first is 5 more than two times the second.

This problem is different, you say. It sure is! Since statement (*c*) tells you to do something with the numbers, you have to represent them first. So, let's see another way to use a total or sum.

Let $\quad x$ = smaller number
$\quad 50 - x$ = larger number

The use of the *total minus x* to represent a second unknown occurs quite often in word problems. It is like saying, "The sum of two numbers is 10. One number is 6. What is the other number?" You know that the other number is 4 because you said to yourself $10 - 6 = 4$. That is why if the sum is 50 and one number is x, we subtract x from 50 and let $50 - x$ represent the second number or unknown. If you know the sum (total), you can subtract one part (x) to get the other part.

4. Let's get back to the problem.
   (*a*) "Two numbers whose sum is 50"
   Let $\quad x$ = first number
   $\quad 50 - x$ = second number

   (*b*) "Three times the first" can be written as

   $$3x$$

   (*c*) "Is" can be written as

   $$=$$

   (Remember that "is, was, will be" are the equals sign.)
   (*d*) "Five more than" is

   $$+5$$

   (*e*) "Twice the second" is

   $$2(50 - x)$$

3

5. Now put it together:

Equation:

$$3x = +5 + 2(50 - x)$$

Only it is better to say

$$3x = 2(50 - x) + 5$$

When you have a problem stating "5 less than," write the -5 on the right of the expression. For example, "5 less than x" translates as $x - 5$. It's good to be consistent and put the "more than" and "less than" quantities on the right: $x + 5$ and $x - 5$; not $+5 + x$ and $-5 + x$.

Solution

$$3x = 2(50 - x) + 5$$

$$3x = 100 - 2x + 5$$

$$5x = 105$$

$$\left. \begin{matrix} x = 21 \\ 50 - x = 29 \end{matrix} \right\} \quad Answers$$

Check: The sum of the numbers is 50. Hence $21 + 29 = 50$ and substituting x in the equation $3x = 2(50 - x) + 5$,

$$3(21) = 2(50 - 21) + 5$$

$$63 = 100 - 42 + 5$$

$$63 = 63$$

EXAMPLE 3

Separate 71 into two parts such that one part exceeds the other by 7. (*Hint:* "Separate 71 into two parts" means two numbers have a sum of 71. "Such that" means in order to have or so that.)

1. Read the problem.
2. What is the question (what are you looking for)? The problem really tells you to find two numbers whose sum is 71.
3. Separate 71 into two parts.

    Let $\quad x =$ smaller part
    $71 - x =$ larger part (this is the "total minus x" concept again)

4. One part exceeds the other by 7. The larger minus the smaller equals the difference, and the larger exceeds the smaller by 7.

$$(71 - x) - x = 7$$

("Exceeds" translates as a minus sign and "by" as an equals sign.)

Solution

$$(71 - x) - x = 7$$
$$71 - 2x = 7$$
$$-2x = 7 - 71$$
$$-2x = -64$$
$$x = 32$$
$$71 - x = 39$$

$\left.\right\}$ *Answers*

Now someone may say, "But I did it differently and got the same answer." There are many different ways to do *some* problems, all of them correct. If we showed you all of them, you would be utterly confused, so if your method *always* works, you are probably correct. It is like saying:

$$8 - 6 = 2$$
$$8 = 2 + 6$$
$$8 - 2 = 6$$

All three are correct statements. We prefer $8 - 6 = 2$ because it "translates" exactly. The words translate into symbols in the same order as you read the statement.

8 exceeds 6 by 2

$$8 \quad - \quad 6 = 2$$

Here is one alternate solution.

Let x = smaller part
$x + 7$ = larger part

Then

$$x + (x + 7) = 71$$
$$2x + 7 = 71$$
$$2x = 64$$
$$x = 32$$
$$x + 7 = 39$$

$\left. \right\}$ *Answers*

Consecutive Integer Problems

There are some special kinds of numbers often found in algebra problems. These are consecutive integers. They are generally qualified as positive. They may be

1. **Consecutive integers.** Consecutive integers are, for example, 21, 22, 23. The difference between consecutive integers is 1.

   Example:
   Represent three consecutive integers.
   Let x = first consecutive integer
   $x + 1$ = second consecutive integer
   $x + 2$ = third consecutive integer

2. **Consecutive even integers.** Consecutive even integers might be 2, 4, 6. The difference between consecutive *even* integers is 2. The first integer in the sequence has to be even, of course.

Example:

Represent three consecutive even integers.

Let x = first even integer

    $x + 2$ = second even integer

    $x + 4$ = third even integer

3. **Consecutive odd integers.** An example of consecutive *odd* integers is 5, 7, 9. Here we find the difference is also 2. But the first integer is *odd*.

Example:

Let x = first consecutive odd integer

    $x + 2$ = second consecutive odd integer

    $x + 4$ = third consecutive odd integer

Note that both the *even* integer problem and the *odd* integer problem are set up exactly the same. The difference is that *x* represents an *even* integer in one and an *odd* integer in the other.

EXAMPLE 4

Find three consecutive integers whose sum is 87.

Solution

Let x = first consecutive integer   (smallest number)

    $x + 1$ = second consecutive integer

    $x + 2$ = third consecutive integer

Equation:

$$x + (x + 1) + (x + 2) = 87$$

$$3x + 3 = 87$$

$$3x = 84$$

$$\left. \begin{array}{r} x = 28 \\ x + 1 = 29 \\ x + 2 = 30 \end{array} \right\} \quad Answers$$

Check: The numbers 28, 29, and 30 are consecutive integers. The sum of 28, 29, and 30 is 87.

EXAMPLE 5

Find three consecutive *even* integers such that the largest is three times the smallest.

Solution

Let x = first consecutive even integer
$x + 2$ = second consecutive even integer
$x + 4$ = third consecutive even integer

Equation: The largest is three times the smallest.

$$x + 4 = 3x$$

$$-2x = -4$$

$$x = 2$$

$$x + 2 = 4 \; \bigg\} \quad Answers$$

$$x + 4 = 6$$

Check: The numbers 2, 4, 6 are even, consecutive integers. The largest is three times the smallest. Thus, $6 = 2(3)$. Always be sure that *all* the unknowns at the beginning of the problem are solved for at the end.

EXAMPLE 6

Four consecutive *odd* integers have a sum of 64. Find the integers.

Solution

Let x = first consecutive odd integer
$x + 2$ = second consecutive odd integer
$x + 4$ = third consecutive odd integer
$x + 6$ = fourth consecutive odd integer

Equation: The first integer + second integer + third integer + fourth integer = sum.

8

$$x + (x + 2) + (x + 4) + (x + 6) = 64$$

$$4x + 12 = 64$$

$$4x = 52$$

$$\left.\begin{aligned} x &= 13 \\ x + 2 &= 15 \\ x + 4 &= 17 \\ x + 6 &= 19 \end{aligned}\right\} \quad \textit{Answers}$$

Check: The sum of 13, 15, 17, and 19 is 64.

Now it is time to try some problems by yourself. Remember:

1. Read the problem all the way through.
2. Find out what you are trying to solve for.
3. Set up the unknown with "Let $x =$" starting every problem. If possible, let x represent the smallest unknown.
4. Read the problem again, a small step at a time, translating words into algebraic statements to set up the unknowns and to state the equation.

Supplementary Number Problems

1. There is a number such that three times the number minus 6 is equal to 45. Find the number.
2. The sum of two numbers is 41. The larger number is 1 less than twice the smaller number. Find the numbers.
3. Separate 90 into two parts so that one part is four times the other part.
4. The sum of three consecutive integers is 54. Find the integers.
5. There are two numbers whose sum is 53. Three times the smaller number is equal to 19 more than the larger number. What are the numbers?

9

6. There are three consecutive odd integers. Three times the largest is seven times the smallest. What are the integers?

7. The sum of four consecutive even integers is 44. What are the numbers?

8. There are three consecutive integers. The sum of the first two is 35 more than the third. Find the integers.

9. A 25-foot-long board is to be cut into two parts. The longer part is 1 foot more than twice the shorter part. How long is each part?

10. Mrs. Mahoney went shopping for some canned goods which were on sale. She bought three times as many cans of tomatoes as cans of peaches and two times as many cans of tuna as cans of peaches. If Mrs. Mahoney purchased a total of 24 cans, how many of each did she buy?

11. The first side of a triangle is 2 inches shorter than the second side. The third side is 5 inches longer than the second. If the perimeter of the triangle is 33 inches, how long is each side?

12. In the afternoon, Kerrie and Shelly rode their bicycles 4 miles more than three times the distance in miles they rode in the morning on a trip to the lake. If the entire trip was 112 miles, how far did they ride in the morning and how far in the afternoon?

13. Mr. and Mrs. Patton and their daughter Carolyn own three cars. Carolyn drives 10 miles per week farther with her car than her father does with his. Mr. Patton drives twice as many miles per week as Mrs. Patton. If their total mileage per week is 160 miles, how many miles per week does each drive?

14. There were 104,830 people who attended a rock festival. If there were 8110 more boys than girls, and 24,810 fewer adults over 50 years of age than there were girls, how many of each group attended the festival?

15. In a 3-digit number, the hundreds digit is 4 more than the units digit and the tens digit is twice the hundreds digit. If the sum of the digits is 12, find the 3 digits. Write the number.

Solutions to Supplementary Number Problems

1. Let x = number
   Three times the number minus 6 is 45.

$$3x - 6 = 45$$

$$3x = 51$$

$$x = 17 \qquad Answer$$

2. Let x = smaller number
$41 - x$ = larger number   (total minus x)

The larger number is twice the smaller number less 1.

$$41 - x = 2x - 1$$
$$-3x = -42$$
$$\left.\begin{array}{l} x = 14 \\ 41 - x = 27 \end{array}\right\} \text{ \textit{Answers}}$$

Alternate Solution

Let x = smaller number
$2x - 1$ = larger number   (larger number is 1 less than twice the smaller)

The sum of the two numbers is 41.

$$x + (2x - 1) = 41$$
$$3x = 42$$
$$\left.\begin{array}{l} x = 14 \\ 2x - 1 = 27 \end{array}\right\} \text{ \textit{Answers}}$$

3. Let x = smaller part
$90 - x$ = larger part   (total minus x)

The larger part is four times the smaller part.

$$90 - x = 4x$$
$$-5x = -90$$
$$\left.\begin{array}{l} x = 18 \\ 90 - x = 72 \end{array}\right\} \text{ \textit{Answers}}$$

Alternate Solution

Let x = smaller part
$4x$ = larger part   (larger part is four times the smaller)

The sum of the two numbers is 90.

$$x + 4x = 90$$

$$5x = 90$$

$$x = 18 \left.\begin{array}{c} \\ \\ \end{array}\right\} \text{Answers}$$

$$4x = 72$$

4. Let $x =$ first consecutive integer
$x + 1 =$ second consecutive integer
$x + 2 =$ third consecutive integer

The sum of the three consecutive integers is 54.

$$x + (x + 1) + (x + 2) = 54$$

$$3x + 3 = 54$$

$$3x = 51$$

$$x = 17$$

$$x + 1 = 18 \quad \text{Answers}$$

$$x + 2 = 19$$

5. Let $x =$ smaller number
$53 - x =$ larger number $\qquad$ (total minus x)

Three times the smaller is the larger plus 19.

$$3x = (53 - x) + 19$$

$$4x = 72$$

$$x = 18$$

$$53 - x = 35 \quad \text{Answers}$$

Alternate Solution

Let $x =$ smaller number
$3x - 19 =$ larger number $\quad$ (larger is 19 less than three times the smaller)

The sum is 53.

$$x + 3x - 19 = 53$$

$$4x - 19 = 53$$

$$4x = 72$$

$$\left.\begin{array}{l} x = 18 \\ 3x - 19 = 35 \end{array}\right\} \text{Answers}$$

6. Let x = first consecutive odd integer
$x + 2$ = second consecutive odd integer
$x + 4$ = third consecutive odd integer

The difference between consecutive odd integers is 2. Three times the largest is seven times the smallest.

$$3(x + 4) = 7x$$

$$3x + 12 = 7x$$

$$-4x = -12$$

$$\left.\begin{array}{l} x = 3 \\ x + 2 = 5 \\ x + 4 = 7 \end{array}\right\} \text{Answers}$$

7. Let x = first consecutive even integer
$x + 2$ = second consecutive even integer
$x + 4$ = third consecutive even integer
$x + 6$ = fourth consecutive even integer

The difference between consecutive even integers is 2. The sum is 44.

$$x + (x + 2) + (x + 4) + (x + 6) = 44$$

$$4x + 12 = 44$$

$$4x = 32$$

$$x = 8$$

$$\left.\begin{array}{l} x + 2 = 10 \\ x + 4 = 12 \\ x + 6 = 14 \end{array}\right\} \text{Answers}$$

8. Let $\quad x =$ first consecutive integer
   $\quad\quad x + 1 =$ second consecutive integer
   $\quad\quad x + 2 =$ third consecutive integer

The difference between consecutive integers is 1. The sum of the first two is the third plus 35.

$$x + (x + 1) = (x + 2) + 35$$

$$2x + 1 = x + 37$$

$$x = 36$$

$$x + 1 = 37 \quad \Big\} \quad Answers$$

$$x + 2 = 38$$

9. Let $\quad x =$ shorter part
   $\quad 25 - x =$ longer part

$$25 - x = 2x + 1$$

$$-3x = -24$$

$$x = 8$$

$$25 - x = 17 \Big\} \quad Answers$$

Alternate Solution

Let $\quad x =$ shorter part
$\quad 2x + 1 =$ longer part

$$x + 2x + 1 = 25$$

$$3x = 24$$

$$x = 8$$

$$2x + 1 = 17 \Big\} \quad Answers$$

Check: The sum is 25 and $8 + 17 = 25$.

10. Let $x =$ number of cans of peaches $\quad$ (smallest number)
   $\quad 3x =$ number of cans of tomatoes
   $\quad 2x =$ number of cans of tuna

$$x + 3x + 2x = 24$$

$$6x = 24$$

$$x = 4$$

$$3x = 12 \biggr\} \quad Answers$$

$$2x = 8$$

Check: The sum is 24 and $4 + 12 + 8 = 24$.

11. Let $\qquad x =$ length of first side in inches $\qquad$ (shortest side)
$\qquad x + 2 =$ length of second side in inches
$\qquad x + 2 + 5 =$ length of third side in inches

$$x + (x + 2) + (x + 2 + 5) = 33$$

$$3x + 9 = 33$$

$$3x = 24$$

$$x = 8$$

$$x + 2 = 10 \biggr\} \quad Answers$$

$$x + 2 + 5 = 15$$

Check: The sum is 33 and $8 + 10 + 15 = 33$.

12. Let $\quad x =$ number of miles in the morning $\quad$ (smaller distance)
$\qquad 3x + 4 =$ number of miles in the afternoon

$$x + 3x + 4 = 112$$

$$4x + 4 = 112$$

$$4x = 108$$

$$x = 27$$

$$3x + 4 = 85 \biggr\} \quad Answers$$

Check: The sum is 112 and $27 + 85 = 112$.

13. Let x = number of miles Mrs. Patton drives   (smallest number)

$2x$ = number of miles Mr. Patton drives

$2x + 10$ = number of miles Carolyn drives

$$x + 2x + (2x + 10) = 160$$

$$5x + 10 = 160$$

$$5x = 150$$

$$x = 30$$

$$2x = 60 \quad \text{Answers}$$

$$2x + 10 = 70$$

Check: The sum is 160 and $30 + 60 + 70 = 160$.

14. Let x = number of girls   (smallest group)

$x + 8110$ = number of boys

$x - 24810$ = number of adults

$$x + x + 8110 + x - 24810 = 104{,}830$$

$$3x - 16700 = 104{,}830$$

$$3x = 121{,}530$$

$$x = 40{,}510$$

$$x + 8110 = 48{,}620 \quad \text{Answers}$$

$$x - 24{,}810 = 15{,}700$$

Check: The total number of people is 104,830 and $40{,}510 + 48{,}620 + 15{,}700 = 104{,}830$.

15. Let x = units digit   (smallest number)

$x + 4$ = hundreds digit

$2(x + 4)$ = tens digit

$$x + (x + 4) + 2(x + 4) = 12$$
$$x + x + 4 + 2x + 8 = 12$$
$$4x = 0$$
$$\left.\begin{array}{r} x = 0 \\ x + 4 = 4 \\ 2(x + 4) = 8 \end{array}\right\} \text{ Answers}$$

The number is 480.

Check: The sum of the digits is 12 and $0 + 4 + 8 = 12$.

Chapter 2

Time, Rate, and Distance

Now that you know how to tackle a word problem and can work the number problems, let's try a new type of problem. Because most of you understand the time, rate, and distance relationship, let's learn a method for setting up time, rate, and distance problems.

First, a little review. If you travel for 2 hours at 50 mph to reach a destination, you know (I hope) that you would have traveled 100 miles. In short, time multiplied by rate equals distance or $t \times r = d$. It is convenient to have a diagram in time, rate, and distance problems. There are usually two moving objects. (Sometimes there is one moving object traveling at two different speeds at different times.) Show in a small sketch the direction and distance of each movement. Then put the information in a simple diagram.

For example, one train leaves Chicago for Boston and at the same time another train leaves Boston for Chicago on the same track but traveling at a different speed. The trains travel until they meet. Your sketch would look like this:

Chicago —— First train ——▶ ◀—— Second train —— Boston

Note that *times* are equal, but distances are unequal.

Or suppose two trains leave the same station at different times traveling in the same direction. One overtakes the other. The distances are equal at the point where one overtakes the other.

Mr. Nemuth leaves New York for a drive to Long Island. Later he returns to New York.

New York   Long Island   Distances are equal

The diagram for your time, rate, and distance information might look like this:

	Time	Rate	Distance
First object			
Second object			

You should fill in the appropriate information from each problem. Rate is in miles per hour (mph) and time is in hours. Distance will therefore be in miles. The basic relationship is *time multiplied by rate equals distance.*

EXAMPLE I

A freight train starts from Los Angeles and heads for Chicago at 40 mph. Two hours later a passenger train leaves the same station for Chicago traveling at 60 mph. How long will it be before the passenger train overtakes the freight train?

Steps

1. Read the problem through, carefully.
2. The question at the end of the problem asks "how long?" (which means time) for the passenger train. This question is your unknown.

Solution

First, draw a sketch of the movement:

Notice that distances are equal.

Second, make a diagram to put in information:

	Time	Rate	Distance
Freight train			
Passenger train			

Now, read the problem again from the beginning for those individual steps. It says the freight train traveled at 40 mph, so fill in the box for the rate of the freight train:

	Time	Rate	Distance
Freight train		40	
Passenger train			

Continue reading: "Two hours later a passenger train . . . traveling at 60 mph." Fill this figure in. You will always know either *both rates* or *both times* in the beginning problems.

	Time	Rate	Distance
Freight train		40	
Passenger train		60	

But so far you haven't any unknown. The question asked is, "How long before the passenger train overtakes the freight train?" As you know by now, you usually put the unknown down first. But when you use the time, rate, and distance table, it helps to put down known facts first because the problem tells you either both the times or both the rates.

Now put in *x* for the time for the passenger train. It's the unknown to be solved for. By the way, did you notice by your sketch that the *distances are equal* when one train overtakes the other? This fact is very important to remember.

Let x = time in hours for the passenger train

	Time	Rate	Distance
Freight train		40	
Passenger train	x	60	

Don't put anything in the distance box until all other information is filled in. Fill in both rates or both times. The other two will be filled in as unknowns. You will multiply time by rate to get distance and put this product in the distance box to represent distance. (Later you may have problems where *each* distance is given, but this type of problem results in fractions. For now, the distances will not be given.) Back to the problem. You have filled in both rates and one time. You must represent the time for the freight train in terms of *x* because you know both rates, and therefore both times are unknown. The problem stated that the passenger train started 2 hours after the freight train, so the freight train took 2 hours longer. You can represent the time for the freight train by $x + 2$.

	Time	Rate	Distance
Freight train	$x + 2$	40	
Passenger train	x	60	

Now time multiplied by rate equals distance ($t \times r = d$), so multiply what you have in the time box by what you have in the rate box and put the result in the distance box:

	Time	Rate	Distance
Freight train	$x + 2$	40	$40(x + 2)$
Passenger train	x	60	$60x$

Every time, rate, and distance problem has some kind of relationship between the distances. This one had the distances equal. That is, the trains traveled the same distance because they started at the same place and traveled until one caught up with the other. This fact was not stated. You have to watch for the relationship. We have two distances in the table above. $40(x + 2)$ represents the distance for the freight train. $60x$ represents the distance for the passenger train. Set these two distances equal for your equation:

$$40(x + 2) = 60x$$

$$40x + 80 = 60x$$

$$-20x = -80$$

$$x = 4 \text{ hours} \qquad Answer$$

Check:

$$40(4 + 2) = 60(4)$$

$$240 = 240$$

Some words of explanation. There was only *one* unknown asked for, but you had to use two to work the problem. Also, you could have used x for the freight train's time and $x - 2$ for the passenger train's time. Always be very careful that you have *answered the question.*

There are several types of time, rate, and distance problems. We have seen one object overtake another, starting from the same place at a different time. Here is another type.

EXAMPLE 2

A car leaves San Francisco for Los Angeles traveling an average of 70 mph. At the same time, another car leaves Los Angeles for San Francisco traveling 60 mph. If it is 520 miles between San Francisco and Los Angeles, how long before the two cars meet, assuming that each maintains its average speed?

Steps

1. Find out if you know both speeds or both times. In this problem we find that both speeds are given.
2. What is the unknown? The question is, "How long before the two cars meet?" So their time is unknown. But which shall we let x equal? *They leave at the same time* and *meet at the same time*, so the traveling times must be equal if neither one stops. Watch for the times being equal when two objects start at the same time and meet at the same time.
3. What fact is known about distance? The 520 miles is the *total* distance, so here is equation information. Remember, don't put distance information in the distance box unless you have to (for example, when the exact distance each object travels is given).

Solution

First, draw a sketch of movement:

Los Angeles ——————▶ ◀—————— San Francisco

Second, make a diagram and fill in all the information you have determined. And remember, time multiplied by rate equals distance in the table.

Let x = time in hours for the two cars to meet

	Time	Rate	Distance
Car from Los Angeles	x	60	$60x$
Car from San Francisco	x	70	$70x$

You know that the total distance is 520 miles, so add the two distances in the diagram and set them equal to 520 miles for the equation.

$$60x + 70x = 520$$
$$130x = 520$$
$$x = 4 \text{ hours} \qquad Answer$$

Check:

$$60(4) + 70(4) = 520$$
$$520 = 520$$

EXAMPLE 3

Two planes leave New York at 10 A.M., one heading for Europe at 600 mph and one heading in the opposite direction at 150 mph. (So it isn't a jet!) At what time will they be 900 miles apart? How far has each traveled?

Steps

1. You know both *speeds.*
2. The times must be unknowns and they are equal (or the same) because the planes leave at the same time and travel until a certain given time (when they are 900 miles apart).
3. The total distance is 900 miles.

Warning! Do not let $x = clock$ time, that is, the time of day. It stands for *traveling* time in hours! The rate is in miles per hour, so time must be in hours in all time, rate, and distance problems.

Solution
Sketch:

New York

Slow plane | Fast plane

Let x = time in hours for planes to fly 900 miles apart

Remember, x equals the time for both the planes, since their times are the same in this problem.

Diagram:

	Time	Rate	Distance
Fast plane	x	600	$600x$
Slow plane	x	150	$150x$

Equation:

$$600x + 150x = 900$$
$$750x = 900$$
$$x = \frac{900}{750}$$
$$x = 1\frac{1}{5} \text{ hours} \qquad Answer$$

Is that what the problem asked for? Always check the question, especially on time, rate, and distance problems. The problem asked *what time* they would be 900 miles apart. Since the problem says they left at 10 A.M., add $1\frac{1}{5}$ hours to that time. So the answer is 11:12 A.M.

The problem also asked how far each would travel. Since we know that rate multiplied by time equals distance,

$$600x = \text{distance for fast plane}$$

and substituting the value of x,

$$600\left(1\frac{1}{5}\right) = 600\left(\frac{6}{5}\right) = 720 \text{ miles} \qquad Answer$$

26

$$150x = \text{distance for slow plane}$$
$$150\left(1\frac{1}{5}\right) = 150\left(\frac{6}{5}\right) = 180 \text{ miles} \quad Answer$$

Check:

$$720 + 180 = 900 \text{ miles}$$

EXAMPLE 4

Mr. Derbyshire makes a business trip from his house to Loganville in 2 hours. One hour later, he returns home in traffic at a rate 20 mph less than his rate going. If Mr. Derbyshire is gone a total of 6 hours, how fast did he travel on each leg of the trip?

Steps

1. This problem does not give us the rates, but it does give us the time both ways. It gives the time to Loganville as 2 hours. The total time is 6 hours, but he did not travel during one of those hours. Deducting this hour leaves only 5 hours of *total traveling time*. That means that if it took 2 hours to travel to Loganville, it must have taken 3 hours to travel back.
2. The rates are both unknown. We can let x equal either one. Just be careful that the faster rate goes with the shorter time. If x equals the rate going, then $x - 20$ equals the rate returning. If x equals the rate returning, then $x + 20$ equals the rate going. Either way is correct.

Solution
Sketch:

Let $\quad x = $ rate in mph going to Loganville
$x - 20 = $ rate in mph returning home

Diagram:

	Time	Rate	Distance
Toward Loganville	2	x	$2x$
Toward home	3	$x - 20$	$3(x - 20)$

Equation: The distance to Loganville equals the distance back home.

$$2x = 3(x - 20)$$

$$-x = -60$$

$$x = 60 \text{ mph} \quad \text{(rate going)}$$

$$x - 20 = 40 \text{ mph} \quad \text{(rate returning)} \Bigg\} \; Answers$$

Check:

$$2(60) = 3(60 - 20)$$

$$120 = 3(40)$$

$$120 = 120$$

EXAMPLE 5

Jake and Jerry went on a camping trip with their motorcycles. One day Jerry left camp on his motorcycle to go to the village. Ten minutes later Jake decided to go too. If Jerry was traveling 30 mph and Jake traveled 35 mph, how long was it before Jake caught up with Jerry?

Steps

1. We know both speeds.
2. We know that Jake traveled for 10 minutes *less* than Jerry. But minutes cannot be used in time, rate, and distance problems because rate is in miles per *hour*. So we must change 10 minutes into 10/60 or one-sixth of an hour.
3. They travel the same distance so the distances are equal.

28

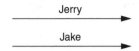

Let x = time in hours for Jake to catch Jerry

Diagram:

	Time	Rate	Distance
Jerry	$x + \dfrac{1}{6}$	30	$30\left(x + \dfrac{1}{6}\right)$
Jake	x	35	$35x$

Equation:

$$35x = 30\left(x + \frac{1}{6}\right)$$
$$35x = 30x + 5$$
$$5x = 5$$
$$x = 1 \text{ hour (Jake's time)} \qquad \textit{Answer}$$

Alternate Solution

	Time	Rate	Distance
Jerry	x	30	$30x$
Jake	$x - \dfrac{1}{6}$	35	$35\left(x - \dfrac{1}{6}\right)$

If you use the alternate solution, you should be good at fractions! Also, be sure to answer the question and solve for $x - \dfrac{1}{6}$, Jake's time.

Note: Can you guess why this problem is here? You must be very careful to have time always in hours or fractions of an hour because rate is in miles per hour.

Basically, our time, rate, and distance problems look like the following:

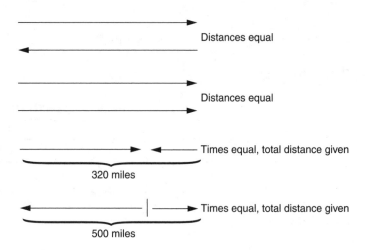

But some time, rate, and distance problems have tricky variations.

EXAMPLE 6

Two cars are headed for Las Vegas. One is 50 miles ahead of the other on the same road. The one in front is traveling 60 mph while the second car is traveling 70 mph. How long will it be before the second car overtakes the first car?

Steps

1. The speeds are given.
2. The times are the unknowns and are equal.
3. The distances are not equal; this inequality is the tricky variation we mentioned above. But if we add 50 miles to the distance of one, it will equal the distance of the other.

Solution
Sketch:

50 miles

First car

Second car

Let x = time in hours for second car to overtake first

Diagram:

	Time	Rate	Distance
First car	x	60	$60x$
Second car	x	70	$70x$

Equation:

$$60x + 50 = 70x$$
$$-10x = -50$$
$$x = 5 \text{ hours} \quad \textit{Answer}$$

If one car is 50 miles ahead of the other at the same time, it will take the cars the same length of time to reach the spot where they are together (or where one overtakes the other). So each takes x hours. The difficulty is that the two cars do not start from the same point, so their distances are not equal. The sketch shows that you have to add 50 miles to the distance the first car travels to make it equal to the distance the second car travels. Of course, the equation could be:

$$70x - 50 = 60x \quad \text{or} \quad 70x - 60x = 50$$

Any one of the three equations gives the same result, and any one is correct. If you can see the relationship better one way than another, always do the problem that way. Again it is like

$$6 + 2 = 8$$
$$8 - 2 = 6$$
$$8 - 6 = 2$$

All are correct statements of the relationship.

Time, Rate, and Distance Problems Involving Moving Air (Wind) or Moving Water (Current)

Some more difficult problems have planes flying in a wind or boats traveling in moving water. The only problems of this type which we can solve are those where the objects move directly *with* or *against* the wind or water. The plane must have a direct headwind or tailwind, and the boat must be going upstream or downstream. In this type of problem the plane's speed in still air would be increased by a tailwind or decreased by a headwind to determine how fast it actually covers the ground. For example, a plane flies 200 mph in still air. This is called airspeed. If there is a 20-mph headwind blowing, it would decrease the speed over the ground by 20 mph, so the ground speed of the plane would be 200 – 20 or 180 mph. The *ground speed* is the *rate* in time, rate, and distance problems. A headwind *reduces* the speed of the plane by the velocity of the wind. A tailwind *increases* the speed of the plane over the ground by the velocity of the wind. A plane with an airspeed (speed in still air) of 400 mph with a 30-mph tailwind actually travels over the ground (ground speed) at 430 mph. A current affects a boat in the same way.

EXAMPLE 7

A plane takes 5 hours to fly from Los Angeles to Honolulu and $4\frac{1}{11}$ hours to return from Honolulu to Los

Angeles. If the wind velocity is 50 mph from the west on both trips, what is the airspeed of the plane? (Airspeed is the speed of the plane in still air.)

Steps

1. The two times are given.
2. You are asked to find the speed of the plane in *still* air (airspeed).
3. Going to Honolulu you have a *headwind*, so *subtract* the velocity of the wind. Returning to Los Angeles, you have a *tailwind*, so *add* the wind velocity to the airspeed.
4. The distances are equal.

Solution

Sketch: Use map directions.

Let x = speed of plane in still air (airspeed)

$x - 50$ = speed of plane over the ground on trip from Los Angeles to Honolulu

$x + 50$ = speed of plane over the ground on trip from Honolulu to Los Angeles

Ground speed determines how long it takes the plane to travel from one place to another.

Diagram:

	Time	Rate	Distance
To Honolulu against wind	5	$x - 50$	$5(x - 50)$
To Los Angeles with wind	$4\frac{1}{11}$	$x + 50$	$4\frac{1}{11}(x + 50)$

Equation: The distances are equal.

$$5(x - 50) = 4\frac{1}{11}(x + 50)$$

$$5(x - 50) = \frac{45}{11}(x + 50)$$

$$\frac{5x - 250}{1} = \frac{45x + 2250}{11}$$

Multiply both sides by 11 to clear fractions.

$$55x - 2750 = 45x + 2250$$

$$10x = 500$$

$$x = 500 \text{ mph} \quad \text{(airspeed)} \quad \textit{Answer}$$

Check:

$$5(500 - 50) = \frac{45}{11}(500 + 50)$$

$$2250 = \frac{24,750}{11}$$

$$2250 = 2250$$

The plane will fly at the same *airspeed* regardless of the wind velocity. The speed at which it actually covers the ground is the airspeed plus or minus the wind velocity, assuming a direct tailwind or headwind.

EXAMPLE 8

In his motorboat, a man can go downstream in 1 hour less time than he can go upstream the same distance. If the current is 5 mph, how fast can he travel in still water if it takes him 2 hours to travel upstream the given distance?

Steps

1. The times are 2 hours upstream and 1 hour downstream.
2. The rates are unknown. If you let x equal his rate of traveling in still water, his rate upstream will be $x - 5$ and his rate downstream will be $x + 5$. You subtract or add the rate of the current.
3. The distance upstream equals the distance downstream.

Sketch:

Let x = rate of boat in mph in still water
    $x - 5$ = rate of boat in mph upstream
    $x + 5$ = rate of boat in mph downstream

Diagram:

	Time	Rate	Distance
Upstream	2	$x - 5$	$2(x - 5)$
Downstream	1	$x + 5$	$1(x + 5)$

Equation:

$$2(x - 5) = 1(x + 5)$$

$$2x - 10 = x + 5$$

$$x = 15 \text{ mph} \qquad \text{(rate in still water)} \qquad \textit{Answer}$$

Check:

$$2(15 - 5) = 1(15 + 5)$$

$$20 = 20$$

Notes to remember about time, rate, and distance problems:

1. Don't let x equal distance unless you have to. (This will sometimes be necessary when you get to fraction problems when each distance is given.)

2. You will always be given either both rates or both times (or later, both distances).
3. Time has to be in hours or fractions of an hour. You can never use minutes with miles per hour.
4. Check the sketch to see if distances are equal for equation information.
5. If two objects start at the same time, their traveling times are generally equal.

Supplementary Time, Rate, and Distance Problems

1. Sarah leaves Seattle for New York in her car, averaging 80 mph across open country. One hour later a plane leaves Seattle for New York following the same route and flying 400 mph. How long will it be before the plane overtakes the car?

2. A train averaging 50 mph leaves San Francisco at 1 P.M. for Los Angeles 440 miles away. At the same time a second train leaves Los Angeles headed for San Francisco on the same track and traveling at an average rate of 60 mph. At what times does the accident occur?

3. Mr. Kasberg rides his bike at 6 mph to the bus station. He then rides the bus to work, averaging 30 mph. If he spends 20 minutes less time on the bus than on the bike, and the distance from his house to work is 26 miles, what is the distance from his house to the bus station?

4. Two planes leave Kansas City at 1 P.M. Plane A heads east at 450 mph and plane B heads due west at 600 mph. How long will it be before the planes are 2100 miles apart?

5. Doomtown is 200 miles due west of Sagebrush, and Joshua is due west of Doomtown. At 9 A.M. Mr. Archer leaves Sagebrush for Joshua. At 1 P.M. Mr. Sassoon leaves Doomtown for Joshua. If Mr. Sassoon travels at an average speed 20 mph faster than Mr. Archer and they each reach Joshua at 4 P.M., how fast is each traveling?

6. Felicia left Rome at 8 A.M. and drove her Ferrari at 80 mph from Rome to Sorrento. She then took the boat to Capri for the day, returning to Sorrento 5 hours later. On the return trip from

Sorrento to Rome she averaged 60 mph and arrived in Rome at 8 P.M. How far is it from Rome to Sorrento?

7. Robin flies to San Francisco from Santa Barbara in 3 hours. She flies back in 2 hours. If the wind was blowing from the north at a velocity of 40 mph going, but changed to 20 mph from the north returning, what was the airspeed of the plane (its speed in still air)?

8. Timothy leaves home for Skedunk 400 miles away. After 2 hours, he has to reduce his speed by 20 mph due to rain. If he takes 1 hour for lunch and gas and reaches Skedunk 9 hours after he left home, what was his initial speed?

9. A highway patrolman spots a speeding car. He clocks it at 70 mph and takes off after it 0.5 mile behind. If the patrolman travels at an average rate of 90 mph, how long will it be before he overtakes the car?

10. Boy Scouts hiking in the mountains divide into two groups to hike around Lake Sahara. They leave the same place at 10 A.M., and one group of younger boys hikes east around the lake and the older group hikes west. If the younger boys hike at a rate of 3 mph and the older boys hike at a rate of 5 mph, how long before they meet on the other side of the lake if the trail around the lake is 8 miles long? At what time do they meet?

11. A boat has a speed of 15 mph in still water. It travels downstream from Greentown to Glenavon in two-fifths of an hour. It then goes back upstream from Glenavon to Cambria, which is 2 miles downstream from Greentown, in three-fifths of an hour. Find the rate of the current.

12. Carlos can run a mile in 6 minutes. Kevin can run a mile in 8 minutes. If Kevin goes out to practice and starts 1 minute before Carlos starts on his practice run, will Carlos catch up with Kevin before he reaches the mile marker? (Assume a straight track and average speed.)

13. Jeanette rides her bike to the bus station where she barely makes it in time to catch the bus to work. She spends half an hour on her bike and two-thirds of an hour on the bus. If the bus travels 39 mph faster than she travels on her bike, and the total distance from home to work is 40 miles, find the rate of the bike and the rate of the bus.

Solutions to Supplementary Time, Rate, and Distance Problems

I. Sketch:

Car
Seattle ——————→ New York   Distances are equal
Plane

Let x = time in hours for plane to travel same distance as car

Diagram:

	Time	Rate	Distance
Car	$x + 1$	80	$80(x + 1)$
Plane	x	400	$400x$

The plane leaves 1 hour later, so the car travels 1 hour longer.

Equation:

$$80(x + 1) = 400x$$
$$80x + 80 = 400x$$
$$-320x = -80$$
$$x = \frac{80}{320}$$
$$x = \frac{1}{4} \text{ hour} \qquad \textit{Answer}$$

Check:

$$80\left(1 + \frac{1}{4}\right) = 400\left(\frac{1}{4}\right)$$
$$80\left(\frac{5}{4}\right) = 400\left(\frac{1}{4}\right)$$
$$100 = 100$$

Alternate Solution

Let x = time in hours car travels before plane overtakes it

Diagram:

	Time	Rate	Distance
Car	x	80	$80x$
Plane	$x - 1$	400	$400(x - 1)$

The plane takes 1 hour less time than the car.

$$80x = 400(x + 1)$$
$$80x = 400x - 400$$
$$-320x = -400$$
$$x = 1\frac{1}{4} \text{ hours}$$
$$x - 1 = \frac{1}{4} \text{ hour}$$
$\left.\begin{array}{c} \\ \\ \end{array}\right\}$ *Answers*

2. Sketch:

440 miles

San Francisco ——→ ←—— Los Angeles
1 P.M.   1 P.M.

Let x = time in hours for trains to meet

Diagram:

	Time	Rate	Distance
First train	x	50	$50x$
Second train	x	60	$60x$

The trains leave at the same time and arrive at the meeting point at the same time, so the times are the same.

Equation: Distance traveled for the first train + distance traveled for the second train = total distance.

$$50x + 60x = 440$$

$$x = 4 \text{ hours}$$

Since the trains left at 1 P.M., they meet 4 hours later at 5 P.M.   *Answer*

Check:

$$50(4) + 60(4) = 440$$

$$200 + 240 = 440$$

3. **Sketch:**

Let x = time in hours from home to bus station

Diagram:

	Time	Rate	Distance
Bike	x	6	$6x$
Bus	$x - \dfrac{1}{3}$	30	$30\left(x - \dfrac{1}{3}\right)$

Twenty minutes is $\dfrac{1}{3}$ hour.

Equation:

$$6x + 30\left(x - \frac{1}{3}\right) = 26$$

$$6x + 30x - 10 = 26$$

$$x = 1 \text{ hour}$$

40

Since the problem asked the distance from house to bus station, we multiply 6 times x,

$$6x = 6 \text{ miles} \qquad \textit{Answer}$$

Check:

$$6(1) + 30\left(1 - \frac{1}{3}\right) = 26$$
$$6 + 20 = 26$$

4. Sketch:

2100 miles

West ◄———— Plane B   (1 P.M.)   Plane A ———► East

Kansas City

Let x = time in hours each plane travels

Diagram:

	Time	Rate	Distance
Plane A	x	450	$450x$
Plane B	x	600	$600x$

Equation:

$$450x + 600x = 2100$$
$$1050x = 2100$$
$$x = 2 \text{ hours} \qquad \textit{Answer}$$

5. Sketch:

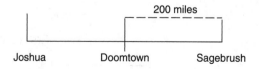

200 miles

Joshua   Doomtown   Sagebrush

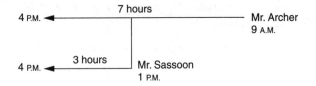

Let x = Mr. Archer's speed in mph
    $x + 20$ = Mr. Sassoon's speed in mph

Diagram:

	Time	Rate	Distance
Archer	7	x	$7x$
Sassoon	3	$x + 20$	$3(x + 20)$

Equation: Mr. Archer travels 200 miles farther than Mr. Sassoon.

$$7x = 3(x + 20) + 200$$

$$7x = 3x + 60 + 200$$

$$4x = 260$$

$$x = 65 \text{ mph} \quad \text{(Mr. Archer's speed)}$$

$$x + 20 = 85 \text{ mph} \quad \text{(Mr. Sassoon's speed)}$$

Answers

Check:

$$7(65) = 3(35) + 200$$

$$455 = 255 + 200$$

$$455 = 455$$

6. Sketch:

Let x = time in hours from Rome to Sorrento

	Time	Rate	Distance
Go	x	80	$80x$
Return	$7 - x$	60	$60(7 - x)$

The *traveling time* is the total time (12 hours) less 5 hours out for Capri, or *7 hours*. The time on the return trip is the total traveling time minus the time on the trip to Sorrento.

Equation:

$$80x = 60(7 - x)$$

$$80x = 420 - 60x$$

$$140x = 420$$

$$x = 3 \text{ hours}$$

Since the problem asked for the distance from Rome to Sorrento, we multiply 80 times x,

$$80x = 80(3) = 240 \text{ miles} \qquad \textit{Answer}$$

Check:

$$80(3) = 60(7 - 3)$$

$$240 = 60(4)$$

$$240 = 240$$

7. Sketch:

San Francisco

Wind going ↓ 40 mph   3 hours   2 hours   Wind returning ↓ 20 mph

Santa Barbara

Let x = speed of plane in still air (airspeed)
           40 = velocity of wind from north (headwind) going
   $x - 40$ = ground speed of plane going to San Francisco (SF)
             against wind
         20 = velocity of wind from north (tailwind) returning
   $x + 20$ = ground speed returning to Santa Barbara (SB)

Diagram:

	Time	Rate	Distance
SB to SF	3	$x - 40$	$3(x - 40)$
SF to SB	2	$x + 20$	$2(x + 20)$

Equation: Distance going equals distance returning.

$$3(x - 40) = 2(x + 20)$$

$$3x - 120 = 2x + 40$$

$$x = 160 \text{ mph} \quad \text{(airspeed)} \quad \textit{Answer}$$

Check:

$$3(160 - 40) = 2(160 + 20)$$

$$360 = 360$$

8. Sketch:

Home ⎯ 2 hours → | ⎯⎯ (Rain) ⎯→ Skedunk
400 miles

Let x = Timothy's original speed in mph

Diagram:

	Time	Rate	Distance
First	2	x	$2x$
Second	6	$x - 20$	$6(x - 20)$

Total time of 9 hours, less 2 hours for the first part, less 1 hour stopped, leaves 6 hours traveling time in the rain.

Equation: First distance + second distance = total distance.

$$2x + 6(x - 20) = 400$$
$$8x - 120 = 400$$
$$8x = 520$$
$$x = 65 \text{ mph} \quad Answer$$

Check:

$$2(65) + 6(65 - 20) = 400$$
$$130 + 270 = 400$$
$$400 = 400$$

9. Sketch:

Let x = time in hours for patrol car to overtake car

Diagram:

	Time	Rate	Distance
Car	x	70	$70x$
Patrol car	x	90	$90x$

Equation: The patrol car travels 0.5 mile farther than the other car.

$$90x = 70x + 0.5$$

$$20x = 0.5$$

Multiply by 10 to clear decimal.

$$200x = 5$$

$$x = \frac{1}{40} \text{ of an hour} \qquad \textit{Answer}$$

If you wish, change the $\frac{1}{40}$ hour to minutes, $\frac{1}{40} \times 60 = 1\frac{1}{2}$ minutes.

Check:

$$90\left(\frac{1}{40}\right) = 70\left(\frac{1}{40}\right) + \frac{1}{2}$$

$$\frac{90}{40} = \frac{70}{40} + \frac{20}{40}$$

$$\frac{90}{40} = \frac{90}{40}$$

10. Sketch:

Let x = time in hours boys take to meet

46

Diagram:

	Time	Rate	Distance
Young scouts	x	3	$3x$
Older scouts	x	5	$5x$

The distance the younger Scouts traveled plus the distance the older Scouts traveled equals the total distance around the lake.

Equation:

$$3x + 5x = 8$$
$$8x = 8$$
$$x = 1 \text{ hour} \quad \textit{Answer}$$

Since the problem also asks what time they meet,

$$10 \text{ A.M.} + 1 \text{ hour} = 11 \text{ A.M.} \quad \textit{Answer}$$

Check:

$$3x + 5x = 8$$
$$3(1) + 5(1) = 8$$
$$3 + 5 = 8$$
$$8 = 8$$

11. **Sketch:**

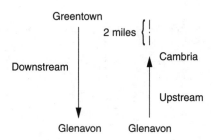

Let x = rate of current in mph
    $15 + x$ = rate of boat downstream in mph
    $15 - x$ = rate of boat upstream in mph

Diagram:

	Time	Rate	Distance
Downstream	$\dfrac{2}{5}$	$15 + x$	$\dfrac{2}{5}(15 + x)$
Upstream	$\dfrac{3}{5}$	$15 - x$	$\dfrac{3}{5}(15 - x)$

Equation: The distance from Greentown to Glenavon is equal to distance from Glenavon to Cambria plus 2 miles.

$$\frac{2}{5}(15 + x) = \frac{3}{5}(15 - x) + 2$$

Remove parentheses.

$$\frac{30 + 2x}{5} = \frac{45 - 3x}{5} + 2$$

Multiply each term by the LCD, 5, to clear fractions.

$$30 + 2x = 40 - 3x + 5(2)$$

$$5x = 25$$

$$x = 5 \text{ mph} \qquad \textit{Answer}$$

Check:

$$\frac{2}{5}(15 + x) = \frac{3}{5}(15 - x) + 2$$

$$\frac{2}{5}(15 + 5) = \frac{3}{5}(15 - 5) + 2$$

$$\frac{2}{5}(20) = \frac{3}{5}(10) + 2$$

$$8 = 6 + 2$$

$$8 = 8$$

12. Sketch:

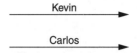

Convert minutes into fractions of an hour first.

$$6 \text{ minutes} = \frac{6}{60} \text{ or } \frac{1}{10} \text{ hour to run 1 mile}$$

$$8 \text{ minutes} = \frac{8}{60} \text{ or } \frac{2}{15} \text{ hour to run 1 mile}$$

Convert the time to run 1 mile into miles per hour by multiplying it by a quantity which turns the time into 1 hour. This quantity is then equivalent to the miles run in 1 hour, or miles per hour.

$$\frac{1}{10} \text{ hour} \times 10 = 1 \text{ hour} \qquad 10 \times 1 \text{ mile} = 10 \text{ mph}$$

$$\frac{2}{15} \text{ hour} \times \frac{15}{2} = 1 \text{ hour} \qquad \frac{15}{2} \times 1 \text{ mile} = 7\frac{1}{2} \text{ mph}$$

Thus Kevin's rate was $7\frac{1}{2}$ mph and Carlos's rate was 10 mph.
Let x = time in hours for Carlos to catch Kevin

Diagram:

	Time	Rate	Distance
Carlos	x	10	$10x$
Kevin	$x + \dfrac{1}{60}$	$\dfrac{15}{2}$	$\dfrac{15}{2}\left(x + \dfrac{1}{60}\right)$

Kevin takes 1 minute more time than Carlos. One minute is $\frac{1}{60}$ of an hour. Distances are equal and unknown in this problem, and we can use this equality for our equation.

$$10x = \frac{15}{2}\left(x + \frac{1}{60}\right)$$

$$10x = \frac{15x}{2} + \frac{1}{8}$$

$$80x = 60x + 1$$

$$20x = 1$$

$$x = \frac{1}{20} \text{ hour}$$

Therefore it takes Carlos $\frac{1}{20}$ hour to catch Kevin. Carlos's distance is $10x$. Substituting the value of x to get distance in miles,

$$10x = 10 \times \frac{1}{20} = \frac{1}{2}$$

Therefore $\frac{1}{2}$ mile from the start Carlos will catch up. Thus, Carlos will catch up with Kevin before he reaches the mile marker.   *Answer*

Check:

$$10x = \frac{15}{2}\left(x + \frac{1}{60}\right)$$

$$10\left(\frac{1}{20}\right) = \frac{15}{2}\left(\frac{1}{20} + \frac{1}{60}\right)$$

$$\frac{1}{2} = \frac{15}{2}\left(\frac{3}{60} + \frac{1}{60}\right)$$

$$\frac{1}{2} = \frac{15}{2}\left(\frac{4}{60}\right)$$

$$\frac{1}{2} = \frac{1}{2}$$

13. Sketch:

Bike	Bus
1/2 hour	2/3 hour

Let x = rate of bicycle in mph
      $x + 39$ = rate of bus in mph

Diagram:

	Time	Rate	Distance
Bike	$\frac{1}{2}$	x	$\frac{1}{2}x$
Bus	$\frac{2}{3}$	$x + 39$	$\frac{2}{3}(x + 39)$

Write $\frac{1}{2}x$ as $\frac{x}{2}$ in your equation.

Equation: Distance on the bike plus the distance on the bus equals the total distance.

$$\frac{x}{2} + \frac{2(x + 39)}{3} = 40$$

Remove parentheses.

$$\frac{x}{2} + \frac{2x + 78}{3} = 40$$

Multiply by the LCD of 6, to clear fractions.

$$6\left(\frac{x}{2}\right) + 6\left(\frac{2x + 78}{3}\right) = 6(40)$$

$$3x + 4x + 156 = 240$$

$$7x = 84$$

$$x = 12 \text{ mph}$$

$$x + 39 = 51 \text{ mph}$$

Answers

Check:

$$\frac{12}{2} + \frac{2(12 + 39)}{3} = 40$$

$$6 + \frac{102}{3} = 40$$

$$6 + 34 = 40$$

$$40 = 40$$

Chapter 3

Mixtures

Now let's tackle a new type of problem—mixture problems. They sound complicated, but we can set them up very methodically. There are two types of mixture problems, those dealing with percent and those involving price. The mechanics of the problems are exactly the same, even though they sound very different. Let us try the percent problems first. There are two important facts in mixture problems involving percents: the *percent* in each mixture and the *amount*. You will be taking two mixtures and dumping them into one pot to make a third mixture. You will always know the percent in each of the two mixtures and also the percent in the total mixture. You will also be given the amount in one of them.

The equation for a mixture problem involving percent when all the percents are given is based on the following fact: The amount in the mixture times the percent of pure stuff equals the *amount* of pure stuff (i.e., pure boric acid, pure silver, pure alcohol).

Example:

Twenty quarts of a mixture containing 15% alcohol would contain

$$20(0.15) = 3 \text{ quarts } pure \text{ alcohol}$$

If you mix two solutions, the amount of pure stuff in the first plus the amount of pure stuff in the second will equal the amount of pure stuff in the total mixture. (The percent of pure stuff in the total mixture is *not* equal to the sum of the other two percents.)

53

Remember these facts:

1. The amount in the first mixture plus the amount in the second mixture equals the amount in the total mixture. We will use this fact to set up unknowns.

2. The amount of pure stuff in the first mixture plus the amount of pure stuff in the second mixture equals the amount of pure stuff in the total mixture. This is the *equation* information. It will look like this:

$$(2\%)(12 \text{ quarts}) + (5\%)(x \text{ quarts}) = (3\%)[(12 + x) \text{ quarts}]$$

or

$$0.02(12) + 0.05(x) = 0.03(12 + x)$$

Percent times amount plus percent times amount equals percent times amount.

Now let's try a problem.

EXAMPLE I

A mixture containing 6% boric acid is to be mixed with 2 quarts of a mixture which is 15% boric acid in order to obtain a solution which is 12% boric acid. How much of the 6% solution must be used?

Steps

1. Draw three diagrams to represent the three solutions. They can be squares, circles, etc. We shall use circles.

2. Find out what the percents are in each original mixture and in the combined mixture and put them in the circles:

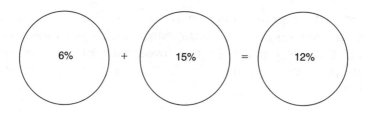

3. Find the *one given amount* and put it in the proper circle:

4. Represent the other two amounts using unknowns. The question asks, "How much of the 6% solution must be used?" so that is what x represents. The total solution will have x quarts plus 2 quarts (or the total of the other two solutions).

Let x = quarts of 6% solution

Solution

You are now ready to solve the problem. If you multiply the amount of solution by the percent of acid in the solution, you will find the amount of *pure* boric acid in each solution. The amount of pure acid in the final solution is equal to the amounts of pure acid in the two original solutions added. This equality gives you the equation information. Keep *all amounts on top* to avoid confusion. Show products of amount times percent in the lower part of the diagram. These products represent the amount of pure acid in each mixture.

Equation:

$$0.06x + 2(0.15) = 0.12(x + 2)$$

It is safer to *eliminate parentheses first* and then eliminate decimals before proceeding to solve for x.

$$0.06x + 0.30 = 0.12x + 0.24$$

To clear decimals, multiply by a power of 10 which will make all numbers whole numbers. Here we multiply by 100 which moves all decimal points *two* places to the right (same as the number of zeros in 100).

$$6x + 30 = 12x + 24$$

$$-6x = -6$$

$$x = 1 \qquad Answer$$

Check:

$$0.06x + 0.30 = 0.12x + 0.24$$

$$0.06(1) + 0.30 = 0.12(1) + 0.24$$

$$0.36 = 0.36$$

EXAMPLE 2

A farmer wants to mix milk containing 3% butterfat with cream containing 30% butterfat to obtain 900 gallons (gal) of milk which is 8% butterfat. How much of each must he use?

Steps

1. Draw the diagram and fill in the known percents and the one known amount. Keep the *amount* on *top* in each diagram.

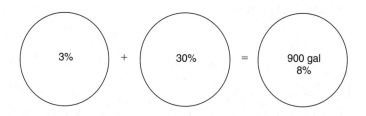

2. Determine what x stands for. Here it will represent the amount of one mixture. There are two mixtures with unknown amounts, and it doesn't make any difference which amount is x.

Let x = gallons of 3% mixture
$900 - x$ = gallons of 30% mixture

3. Represent the amount in the other mixture using x and the amount given in the problem. Remember that the amounts of the two original mixtures add up to the amount of the total mixture. The second unknown is 900

– *x*. (This is another example of the use of *total* – *x* to represent the second unknown.)

4. Multiply each percent by the amount of its mixture and set these products equal:

Solution
Equation:

$$0.03x + 0.30(900 - x) = 0.08(900)$$

Clear parentheses.

$$0.03x + 270 - 0.30x = 72$$

Multiply by 100.

$$3x + 27{,}000 - 30x = 7200$$

$$-27x = -19{,}800$$

$$\left.\begin{array}{r} x = 733\frac{1}{3} \\[2mm] 900 - x = 166\frac{2}{3} \end{array}\right\} \quad \textit{Answers}$$

EXAMPLE 3

Now let's look at a mixture problem involving prices instead of percents. For example, a merchant wishes to mix two grades of coffee, one of which sells for $4.00 per pound and one of which sells for $6.00 per pound. He wants to sell the mixture for $5.50 per pound. If he has 25 pounds of the $4.00 coffee, how much of the $6.00 coffee must he add so that the value of the final mixture is equal to the total value of the other two?

This problem sounds different from the first type, but like the percent problem it has certain basic facts. There are two quantities mixed. There are three *prices* given instead of three percents. One amount is given. The problem can be illustrated by the same type of diagram.

Steps

1. Draw three circles (or squares or rectangles) to represent the three types of coffee, and label the price and the amount in each circle.

Again we must represent the *amount* in each mixture. The question asks, "How much of the $6.00 coffee must be added?" So we let x equal the amount of the $6.00 coffee, and $x + 25$ will be the amount in the total mixture.

Let x = pounds (lb) of $6.00 coffee

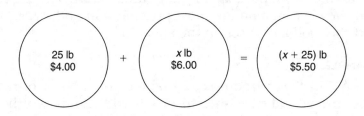

2. The problem states that the *value* of the final mixture is equal to the *total* value of the other two. If we have the

number of pounds and the price per pound, we can multiply to find the value of each mixture.

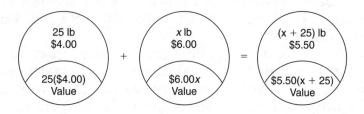

3. The product in the lower part of each circle represents the value of the mixture. The sum of the first two values equals the value of the total.

Solution
Equation:

$$25(4.00) + 6.00x = 5.50(x + 25)$$

$$100 + 6.00x = 5.50x + 137.5$$

Multiply by 10 to clear the decimals. (You will still be correct if you multiply by 100.)

$$1000 + 60x = 55x + 1375$$

$$x = 75 \quad \textit{Answer}$$

Note that in this type of mixture problem the amounts add up to the total amount just as in the first type. And even though the problems are different, the basic procedure is the same. In one we have an amount and a percent. In the second, we have an amount and a price. And in both types, we find the products to set up the equation.

EXAMPLE 4

A Girl Scout troop has 20 pounds of candy worth 80 cents per pound. It wishes to mix it with candy worth 50 cents per pound so that the total mixture can be sold at 60 cents per pound without any gain or loss. How much of the 50-cent candy must be used?

1. Draw a diagram and put in prices and amounts. This problem is set up just like the percent problem except you have *cents* instead of *percent*.

   Let x = pounds of 50-cent candy

2. Multiply the cents per pound times the amounts. This will give you the *value* of each mixture. The value of each original mixture added together should equal the value of the total mixture if there is no gain or loss. (We assume no gain or loss unless the problem states otherwise.)

Remember, 80 cents should be written in decimal form when multiplying (0.80).

Solution
Equation:

$$0.80(20) + 0.50x = 0.60(x + 20)$$

$$16 + 0.50x = 0.60x + 12$$

$$1600 + 50x = 60x + 1200$$

$$50x - 60x = 1200 - 1600$$

$$-10x = -400$$

$$x = 40 \qquad Answer$$

Note: You could clear decimals in this equation by multiplying by 10 instead of 100. It would make all numbers in the equation whole numbers.

Unless otherwise stated, as we said, it is assumed that the value of a total mixture is equal to the sums of the values of the two original mixtures. But you *could* be given other relationships, as for instance, in the following problem.

EXAMPLE 5

A merchant purchased two lots of shoes. One lot he purchased for $32 per pair and the second lot he purchased for $40 per pair. There were 50 pairs in the first lot. How many pairs were in the second lot if he sold them all at $60 per pair and made a gain of $2800 on the entire transaction?

Steps

1. Draw a diagram, fill in all information, and multiply price times quantity to get the total value of each original lot and the value of the total lot.

Let x = pairs in second lot

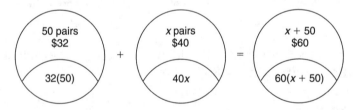

2. Set up the equation by adding $2800 to the sum of the first and second lots. This addition of $2800 equals the gain on the entire transaction.

Solution
Equation:

$$32(50) + 40x + 2800 = 60(x + 50)$$
$$1600 + 40x + 2800 = 60x + 3000$$
$$-20x = -1400$$
$$x = 70 \qquad Answer$$

Check:

$$1600 + 40(70) + 2800 = 60(70 + 50)$$

$$1600 + 2800 + 2800 = 4200 + 3000$$

$$7200 = 7200$$

There are some mixture problems in which water is added to dilute a mixture. Water has 0% acid, of course. You may also add pure ingredients to strengthen a mixture. For example, you might add pure sulphuric acid to a 30% solution to strengthen it. The pure ingredient is 100%. Remember, $0\%(x) = 0$, and $100\%(x) = x$. When the percent form is changed to the decimal form, the decimal is moved two places to the left.

Percent		Decimal
100%	=	1.00
80%	=	0.80
70%	=	0.70

EXAMPLE 6

A chemist needs a solution of tannic acid that is 70% pure. How much distilled water must she add to 5 gallons of acid which is 90% pure to obtain the 70% solution?

Solution

Let x = gallons of water

Diagram:

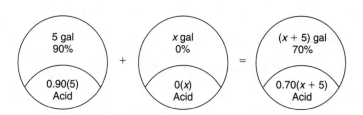

Equation:

$$0.90(5) + 0 = 0.70(x + 5)$$

$$4.5 + 0 = 0.70x + 3.5$$

Multiply by 10.

$$45 = 7x + 35$$

$$-7x = 35 - 45$$

$$-7x = -10$$

$$x = 1\frac{3}{7} \quad \text{Answer}$$

EXAMPLE 7

How much pure alcohol must a nurse add to 10 cubic centimeters (cc) of a 60% alcohol solution to strengthen it to a 90% alcohol solution?

Solution

Let x = cubic centimeters of pure alcohol

Diagram:

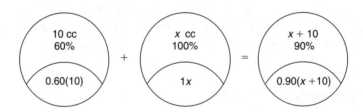

Pure alcohol is 100% alcohol. In decimal form 100% is 1.00 or 1.

Equation:

$$0.60(10) + 1x = 0.90(x + 10)$$

$$6 + x = 0.90x + 9$$

$$60 + 10x = 9x + 90$$

$$x = 30 \qquad \textit{Answer}$$

Check:

$$0.60(10) + 30 = 0.90(30 + 10)$$

$$6 + 30 = 27 + 9$$

$$36 = 36$$

Now that you understand basic mixture problems, we can try a different kind. In this type of problem we take some from, and add some to, a given mixture. (You are more likely to find this type of problem in second-year algebra.)

EXAMPLE 8

A 5-gallon radiator containing a mixture of water and antifreeze was supposed to contain a 50% antifreeze solution. When tested, it was found to have only 40% antifreeze. How much must be drained out and replaced with pure antifreeze so that the radiator will then contain the desired 50% antifreeze solution?

Solution

This time we need another circle for the diagram. The first circle in the diagram shows what the radiator has to start with. The second circle shows what has been drained off. It will always be the *same percent* as the solution from which it is taken. The third circle shows what is put back. It will be the *same amount* as that taken out. The fourth circle represents the final solution and will be the same amount you started with. You then multiply percent times amount to get amount of pure antifreeze in each solution and use the multiples to set up the equation, the same as before.

Let x = gallons of solution drained *and* replaced

Diagram:

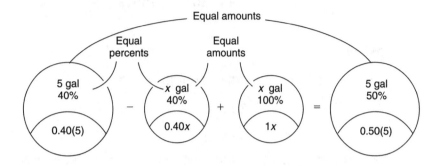

Note that you start with 5 gallons and finish with 5 gallons. The amount you drain off is 40% antifreeze, the same percent as in the original solution.

Equation:

$$0.40(5) - 0.40x + 1x = 0.50(5)$$

$$2.00 - 0.40x + 1x = 2.50$$

$$200 - 40x + 100x = 250$$

$$60x = 50$$

$$x = \frac{5}{6} \quad \textit{Answer}$$

EXAMPLE 9

A chemist has 300 grams of 20% hydrochloric acid solution. He wishes to drain some off and replace it with an 80% solution so as to obtain a 25% solution. How many grams must he drain and replace with the 80% solution?

Solution

Let x = grams of acid to drain and replace

Diagram:

Equation:

$$0.20(300) - 0.20x + 0.80x = 0.25(300)$$
$$60 - 0.20x + 0.80x = 75$$
$$600 - 2x + 8x = 750$$
$$6x = 150$$
$$x = 25 \quad \textit{Answer}$$

Supplementary Mixture Problems

1. A florist wishes to make bouquets of mixed spring flowers. Each bouquet is to be made up of tulips at $30 a bunch and daffodils at $21 a bunch. How many bunches of each should she use to make 15 bunches which she can sell for $24 a bunch?

2. A farmer has 100 gallons of 70% pure disinfectant. He wishes to mix it with disinfectant which is 90% pure in order to obtain 75% pure disinfectant. How much of the 90% pure disinfectant must he use?

3. If an alloy containing 30% silver is mixed with a 55% silver alloy to get 800 pounds of 40% alloy, how much of each must be used?

4. Russ discovers at the end of the summer that his radiator antifreeze solution has dropped below the safe level. If the radiator contains 4 gallons of a 25% solution, how many gallons of pure antifreeze must he add to bring it up to a desired 50% solution? (Assume there is room to add antifreeze without removing any solution.)

5. A store manager wishes to reduce the price on her fresh ground coffee by mixing two grades. If she has 50 pounds of coffee which sells for $10 per pound, how much coffee worth $6 per pound must she mix with it so that she can sell the final mixture for $8.50 per pound?

6. A hospital needs to dilute a 50% boric acid solution to a 10% solution. If it needs 25 liters of the 10% solution, how much of the 50% solution and how much water should it use?

7. Annabell has 25 ounces of a 20% boric acid solution which she wishes to dilute to a 10% solution. How much water does she have to add in order to obtain the 10% solution?

8. Forty liters of a 60% disinfectant solution are to be mixed with a 10% solution to dilute it to a 20% solution. How much of the 10% solution must be used?

9. A doctor orders 20 grams of a 52% solution of a certain medicine. The pharmacist has only bottles of 40% and bottles of 70% solution. How much of each must he use to obtain the 20 grams of the 52% solution?

10. Forty liters of a 60% salt solution are reduced to a 45% solution. How much must be drained off and replaced with distilled water so that the resulting solution will contain only 45% salt?

11. A pharmacist has to fill a prescription calling for 5 ounces of a 12% Argyrol solution. She has 5 ounces of a 15% solution and 5 ounces of a 5% solution. If she starts with the 5 ounces of 15%, how much must she draw off and replace with 5% in order to fill the prescription?

Solutions to Supplementary Mixture Problems

1. Let x = number of $30 bunches

$15 - x$ = number of $21 bunches

Diagram:

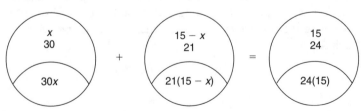

The value of the bunches at $30 each is 30x. There are 15 bunches in total so $15 - x$ is the number of bunches at $21 each. The value of the bunches in the second mixture is $7(15 - x)$. The value of the first mixture plus the value of the second mixture equals the total mixture.

Equation:

$$30x + 21(15 - x) = 24(15)$$

$$30x + 315 - 21x = 360$$

$$9x = 45$$

$$x = 5 \text{ bunches at \$30 each}$$

$$15 - x = 10 \text{ bunches at \$21 each}$$

$\left. \right\}$ *Answers*

Check:

$$30(5) + 21(15 - 5) = 24(15)$$

$$150 + 315 - 105 = 360$$

$$360 = 360$$

2. Let x = gallons of 90% disinfectant

Diagram:

100 gal 70%		x gal 90%		$(x + 100)$ gal 75%
0.70(100)	+	0.90x	=	0.75(x + 100)

Seventy percent of 100 gallons equals pure disinfectant in the first mixture. Ninety percent of x gallons equals pure disinfectant in the second mixture. Seventy-five percent of $(x + 100)$ gallons equals pure disinfectant in the total mixture.

Equation:

$$0.70(100) + 0.90x = 0.75(x + 100)$$
$$70 + 0.90x = 0.75x + 75$$

Multiply by 100.

$$7000 + 90x = 75x + 7500$$
$$15x = 500$$
$$x = 33\frac{1}{3} \qquad Answer$$

Check:

$$0.70(100) + 0.90\left(33\frac{1}{3}\right) = 0.75\left(33\frac{1}{3} + 100\right)$$
$$70 + 30 = 25 + 75$$
$$100 = 100$$

3. Let $\qquad x =$ pounds of 30% alloy
    $800 - x =$ pounds of 55% alloy

Diagram:

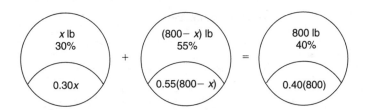

Equation:

$$0.30x + 0.55(800 - x) = 0.40(800)$$

Clear parentheses.

$$0.30x + 440 - 0.55x = 320$$

Multiply by 100.

$$30x + 44,000 - 55x = 32,000$$
$$-25x = -12,000$$

$$\left. \begin{array}{r} x = 480 \\ 800 - x = 320 \end{array} \right\} \quad \textit{Answers}$$

Check:

$$0.30(480) + 0.55(800 - 480) = 0.40(800)$$

$$144 + 440 - 264 = 320$$

$$584 - 264 = 320$$

$$320 = 320$$

4. Let x = gallons of pure antifreeze

Diagram:

Equation:

$$0.25(4) + 1x = 0.50(x + 4)$$

$$1.00 + x = 0.50x + 2.00$$

Multiply by 100.

$$100 + 100x = 50x + 200$$

$$x = 2 \quad \textit{Answer}$$

Check:

$$0.25(4) + 2 = 0.50(2 + 4)$$

$$1 + 2 = 1 + 2$$

5. Let x = pounds of \$6 coffee

Diagram:

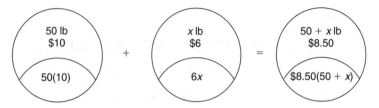

Equation:

$$50(10) + 6x = 8.50(50 + x)$$
$$500 + 6x = 425 + 8.50x$$
$$5000 + 60x = 4250 + 85x$$
$$-15x = -750$$
$$x = 30 \text{ pounds of \$6 coffee} \qquad Answer$$

6. Let x = liters of 50% solution
$25 - x$ = liters of water

Diagram:

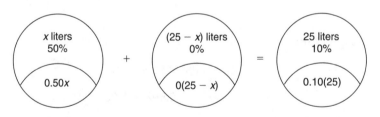

Equation:

$$0.50x + 0 = 0.10(25)$$
$$0.50x = 2.50$$
$$50x = 250$$
$$x = 5$$
$$25 - x = 20 \qquad Answers$$

72

7. Let x = ounces (oz) of water needed

Diagram:

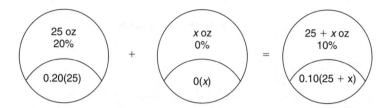

Water has 0% boric acid in it. The percent times the amount in each mixture gives the number of ounces of pure boric acid in each. This amount of pure acid is represented at the bottom of each circle.

Equation:

$$0.20(25) + 0(x) = 0.10(25 + x)$$
$$5 + 0 = 2.5 + 0.1x$$

Multiply by 10 to clear fractions.

$$50 = 25 + x$$
$$x = 25 \quad \textit{Answer}$$

8. Let x = liters of 10% solution

Diagram:

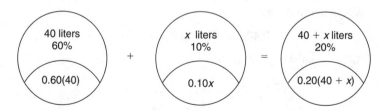

Equation:

$$0.60(40) + 0.10x = 0.20(40 + x)$$
$$24 + 0.10x = 8 + 0.20x$$

Multiply by 10 to clear decimals.

$$240 + 1x = 80 + 2x$$

$$-x = -160$$

$$x = 160 \qquad Answer$$

Check:

$$0.60(40) + 0.10(160) = 0.20(40 + 160)$$

$$24 + 16 = 0.20(200)$$

$$40 = 40$$

9. Let $\qquad x =$ grams of 40% solution
$\qquad 20 - x =$ grams of 70% solution

Diagram:

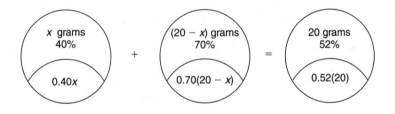

Equation:

$$0.40x + 0.70(20 - x) = 0.52(20)$$

Clear parentheses.

$$0.40x + 14 - 0.70x = 10.4$$

Clear decimals.

$$4x + 140 - 7x = 104$$

$$-3x = -36$$

$$x = 12$$
$$20 - x = 8$$
$\left.\phantom{\begin{matrix}x = 12 \\ 20 - x = 8\end{matrix}}\right\} Answers$

Check:

$$0.40(12) + 0.70(20 - 12) = 0.52(20)$$

$$4.8 + 14 - 8.4 = 10.4$$

$$18.8 - 8.4 = 10.4$$

$$10.4 = 10.4$$

10. Let x = liters of distilled water (and solution drained off)

Diagram:

Equation:

$$0.60(40) - 0.60x + 0(x) = 0.45(40)$$

$$24 - 0.60x = 18$$

$$240 - 6x = 180$$

$$-6x = -60$$

$$x = 10 \quad \textit{Answer}$$

11. Let x = ounces of 5% solution (and ounces of 15% solution drained off)

Diagram:

Equation:

$$0.15(5) - 0.15x + 0.05x = 0.12(5)$$

$$0.75 - 0.15x + 0.05x = 0.60$$

$$75 - 15x + 5x = 60$$

$$-10x = -15$$

$$x = 1\frac{1}{2} \quad \text{Answer}$$

Chapter 4

Coins

Problems about money, either metal or paper, stamps, or any objects of value, are, for want of a better classification, grouped under the heading "Coin Problems." It is very important to identify and label whether it is *how many things* or *how much money* you are representing.

EXAMPLE I

Michael has some coins in his pocket consisting of dimes, nickels, and pennies. He has two more nickels than dimes, and three times as many pennies as nickels. How many of each kind of coin does he have if the total value is 52 cents?

Steps

1. Try to determine which type of coin he has the *fewest of.* This is often a good way to find what to let x stand for. Here he has fewer dimes than nickels or pennies.
2. The question asks *how many* of each kind of coin (not how much they are *worth*). That is, what *number* of each kind of coin does he have? So, let

$$x = \text{number of dimes}$$

3. Go back to one fact at a time. He has two more nickels than dimes.

$$x + 2 = \text{number of nickels}$$

4. The next fact is that he has three times as many pennies as nickels.

$$3(x + 2) = \text{number of pennies}$$

5. The information left for the equation is that the total value is 52 cents. You cannot say that the total *number of coins* equals 52 cents. *Number of* must be changed to *value*. If you had two quarters, you would have 50 cents. You multiplied how many coins by how much each is worth. So let's change our *how many* to *how much*. If you have the number of dimes, multiply by 10 to change to cents. Multiply the number of nickels by 5 to change to cents. The number of pennies is the same as the number of cents.

Number of Coins	**Value in Cents**
x = number of dimes	$10x$ = number of cents in dimes
$x + 2$ = number of nickels	$5(x + 2)$ = number of cents in nickels
$3(x + 2)$ = number of pennies	$3(x + 2)$ = number of cents in pennies

Now we can add the *amounts of money*. If you make it all pennies, there are no decimals.

$$10x + 5(x + 2) + 3(x + 2) = 52$$

$$10x + 5x + 10 + 3x + 6 = 52$$

$$18x = 36$$

$$x = 2$$
$$x + 2 = 4$$
$$3(x + 2) = 12$$

$$\left.\begin{array}{l} x = 2 \\ x + 2 = 4 \\ 3(x + 2) = 12 \end{array}\right\} \text{Answers}$$

Check:

$$2(10 \text{ cents}) = 20 \text{ cents in dimes}$$

$$4(5 \text{ cents}) = 20 \text{ cents in nickels}$$

$$\underline{12(1 \text{ cent}) = 12 \text{ cents in pennies}}$$

$$\text{Total} = 52 \text{ cents}$$

EXAMPLE 2

A coin collector had a collection of silver coins worth $205. There were five times as many quarters as half-dollars (50-cent pieces) and 200 fewer dimes than quarters. How many of each kind of coin did the collector have?

Solution

Number of Coins	Value in Cents
x = number of half-dollars	$50x$ = number of cents in half-dollars
$5x$ = number of quarters	$25(5x)$ = number of cents in quarters
$5x - 200$ = number of dimes	$10(5x - 200)$ = number of cents in dimes

Equation: Remember, everything has been changed to *cents*. So $205 has to be changed to cents by multiplying by 100. $205 = 20,500$ cents.

$$50x + 25(5x) + 10(5x - 200) = 20,500$$
$$50x + 125x + 50x - 2000 = 20,500$$
$$225x = 22,500$$
$$\left.\begin{array}{r} x = 100 \\ 5x = 500 \\ 5x - 200 = 300 \end{array}\right\} \quad Answers$$

Check:

$$100(50 \text{ cents}) = \$\ 50$$
$$500(25 \text{ cents}) = \ 125$$
$$300(10 \text{ cents}) = \underline{\ \ 30}$$
$$\text{Total} = \$205$$

EXAMPLE 3

Mr. Dinkelspiel bought $32.56 worth of stamps. He bought 20 more 19-cent stamps than 50-cent stamps. He bought twice as many 32-cent stamps as 19-cent stamps. How many of each kind of stamp did he buy?

Solution

Number of Stamps	Value in Cents
x = number of 50-cent stamps	$50x$
$x + 20$ = number of 19-cent stamps	$19(x + 20)$
$2(x + 20)$ = number of 32-cent stamps	$32[2(x + 20)]$

Equation: Remember, $32.56 equals 3256 cents.

$$50x + 19(x + 20) + 32[2(x + 20)] = 3256$$
$$50x + 19x + 380 + 64x + 1280 = 3256$$
$$133x = 1596$$
$$x = 12 $$
$$x + 20 = 32 \Big\} \quad Answers$$
$$2(x + 20) = 64 $$

Check:

$$12(50) + 19(32) + 64(32) = 3256$$
$$600 + 608 + 2048 = 3256$$
$$3256 = 3256$$

The principle of converting "how many things" into "how much money" can be applied to many types of problems. Let's try some.

Supplementary Coin Problems

1. A collection of coins has a value of 64 cents. There are two more nickels than dimes and three times as many pennies as dimes. How many of each kind of coin are there?

2. Tanya has ten bills in her wallet. She has a total of $40. If she has one more $5 bill than $10 bills, and two more $1 bills than $5 bills, how many of each does she have? (There are two ways of working this problem. See if you can do it both ways.)

3. Mario bought $21.44 worth of stamps at the post office. He bought 10 more 4-cent stamps than 19-cent stamps. The number of 32-cent stamps was three times the number of 19-cent stamps. He also bought two $1 stamps. How many of each kind of stamp did he purchase?

4. Mr. Abernathy purchases a selection of wrenches for his shop. His bill is $78. He buys the same number of $1.50 and $2.50 wrenches, and half that many of $4 wrenches. The number of $3 wrenches is one more than the number of $4 wrenches. How many of each did he purchase? (*Hint:* If you have not worked with fractions, use decimals for all fractional parts.)

5. A clerk at the Dior Department Store receives $15 in change for her cash drawer at the start of each day. She receives twice as many dimes as fifty-cent pieces, and the same number of quarters as dimes. She has twice as many nickels as dimes and a dollar's worth of pennies. How many of each kind of coin does she receive?

6. A collection of 36 coins consists of nickels, dimes, and quarters. There are three fewer quarters than nickels and six more dimes than quarters. How many of each kind of coin are there?

7. The cash drawer of the market contains $227 in bills. There are six more $5 bills than $10 bills. The number of $1 bills is two more than 24 times the number of $10 bills. How many bills of each kind are there?

8. Jennifer went to the drugstore. She bought a bottle of aspirin and a bottle of Tylenol. The aspirin cost $1.25 more than the Tylenol. She also bought cologne which cost twice as much as the total of the other two combined. How much did each cost if her total (without tax) was $24.75?

9. Terry bought some gum and some candy. The number of packages of gum was one more than the number of mints. The number of

mints was three times the number of candy bars. If gum was 24 cents a package, mints were 10 cents each, and candy bars were 35 cents each, how many of each did he get for $5.72?

10. Jose Ramirez had $50 to buy his groceries. He needed milk at $1.95 a carton, bread at $2.39 a loaf, breakfast cereal at $3.00 a box, and meat at $5.39 a pound. He bought twice as many cartons of milk as loaves of bread and one more package of cereal than loaves of bread. He also bought the same number of pounds of meat as packages of cereal. How many of each item did he purchase if he received $12.25 in change?

Solutions to Supplementary Coin Problems

1.

Number of Coins	**Value in Cents**
x = number of dimes	$10x$ = value of dimes
$x + 2$ = number of nickels	$5(x + 2)$ = value of nickels
$3x$ = number of pennies	$3x$ = value of pennies

The total value of the dimes, nickels, and quarters equals 64 cents.

Equation:

$$10x + 5(x + 2) + 3x = 64$$
$$10x + 5x + 10 + 3x = 64$$
$$18x = 54$$
$$x = 3$$
$$x + 2 = 5 \quad \Big\} \quad Answers$$
$$3x = 9$$

Check:

$$3(10 \text{ cents}) = 30 \text{ cents}$$
$$5(5 \text{ cents}) = 25$$
$$9(1 \text{ cent}) = \underline{9}$$
$$\text{Total} = 64 \text{ cents}$$

82

2.

Number of Bills	Values in Dollars
x = number of $10 bills	$10x$ = value of $10 bills
$x + 1$ = number of $5 bills	$5(x + 1)$ = value of $5 bills
$(x + 1) + 2$ = number of $1 bills	$1(x + 3)$ = value of $1 bills

Equation:

$$10x + 5(x + 1) + (x + 3) = 40$$
$$10x + 5x + 5 + x + 3 = 40$$
$$16x = 32$$
$$x = 2$$
$$x + 1 = 3 \quad \left.\right\} \quad Answers$$
$$x + 3 = 5$$

Check:

$$2(\$10) = \$20$$
$$3(\$5) = \ 15$$
$$5(\$1) = \ \underline{\ 5}$$
$$\text{Total} = \$40$$

Alternate Solution

The information that there are 10 bills was not used in the first solution. If you know *how many* coins, bills, or stamps you have, the problem is simplified because you do not have to convert to money.

Let x = number of $10 bills
  $x + 1$ = number of $5 bills
  $x + 3$ = number of $1 bills

Equation:

$$x + (x + 1) + (x + 3) = 10$$
$$3x + 4 = 10$$

83

$$3x = 6$$

$$\left. \begin{array}{r} x = 2 \\ x + 1 = 3 \\ x + 3 = 5 \end{array} \right\} \quad \textit{Answers}$$

Remember, however, that you are seldom given the *number* of coins.

3.

Number of Stamps	Value in Cents
x = number of 19-cent stamps	$19x$
$x + 10$ = number of 4-cent stamps	$4(x + 10)$
$3x$ = number of 32-cent stamps	$32(3x)$
2 = number of \$1 stamps	$2(100)$

Equation:

$$19x + 4(x + 10) + 32(3x) + 2(100) = 2144$$

$$19x + 4x + 40 + 96x + 200 = 2144$$

$$119x = 1904$$

$$\left. \begin{array}{r} x = 16 \\ x + 10 = 26 \\ 3x = 48 \end{array} \right\} \quad \textit{Answers}$$

Check:

$$19(16) + 4(26) + 32(48) + 200 = 2144$$

$$304 + 104 + 1536 + 200 = 2144$$

$$2144 = 2144$$

4.

Number of Wrenches

x = number of \$4.00 wrenches

$2x$ = number of \$1.50 wrenches

$2x$ = number of \$2.50 wrenches

$x + 1$ = number of \$3.00 wrenches

Value in Dollars

$(x)(4.00) = $ value of $4.00 wrenches

$(2x)(1.50) = $ value of $1.50 wrenches

$(2x)(2.50) = $ value of $2.50 wrenches

$(x + 1)(3.00) = $ value of $3.00 wrenches

Equation:

$$x(4.00) + 2x(1.50) + 2x(2.50) + (x + 1)(3.00) = 78$$

Note in this problem that the decimal is removed when you clear parentheses.

$$4x + 3x + 5x + 3x + 3 = 78$$
$$15x = 75$$
$$x = 5$$
$$2x = 10$$
$$2x = 10 \quad \text{Answers}$$
$$x + 1 = 6$$

Check:

$$5(\$4.00) = \$20$$
$$10(\$1.50) = 15$$
$$10(\$2.50) = 25$$
$$6(\$3.00) = 18$$
$$\text{Total} = \$78$$

5. Number of Coins

$x = $ number of 50-cent pieces

$2x = $ number of 10-cent pieces

$2x = $ number of 25-cent pieces

$4x = $ number of 5-cent pieces

$100 = $ number of 1-cent pieces

Value in Cents

$50x = $ value of 50-cent pieces

$10(2x) = $ value of 10-cent pieces

$25(2x) = $ value of 25-cent pieces

$5(4x) = $ value of 5-cent pieces

$1(100) = $ value of 1-cent pieces

Equation: Fifteen dollars equals 1500 cents.

$$50x + 2x(10) + 2x(25) + 4x(5) + 100 = 1500$$
$$50x + 20x + 50x + 20x + 100 = 1500$$
$$140x = 1400$$

$$x = 10$$
$$2x = 20$$
$$2x = 20$$
$$4x = 40$$

} *Answers*

Check:

$$10(50 \text{ cents}) = \$5$$
$$20(10 \text{ cents}) = 2$$
$$20(25 \text{ cents}) = 5$$
$$40(5 \text{ cents}) = 2$$
$$100(1 \text{ cent}) = \underline{\quad 1}$$
$$\text{Total} = \$15$$

6. Let x = number of quarters (there are *fewer* quarters)
    $x + 3$ = number of nickels
    $x + 6$ = number of dimes

Here we do *not* change to cents because the *number* of coins is given.

Equation:

$$x + (x + 3) + (x + 6) = 36$$
$$3x + 9 = 36$$
$$3x = 27$$
$$x = 9 \text{ quarters}$$
$$x + 3 = 12 \text{ nickels}$$
$$x + 6 = 15 \text{ dimes}$$

} *Answers*

Check:

$$9 + 12 + 15 = 36$$

$$36 = 36$$

7. Let $\quad x = $ number of \$10 bills
$x + 6 = $ number of \$5 bills
$24x + 2 = $ number of \$1 bills

Change all to amount of money in dollars.

$$10x = \text{number of dollars in \$10 bills}$$

$$5(x + 6) = \text{number of dollars in \$5 bills}$$

$$24x + 2 = \text{number of dollars in \$1 bills}$$

Equation:

$$10x + 5(x + 6) + 24x + 2 = 227$$

$$10x + 5x + 30 + 24x + 2 = 227$$

$$39x + 32 = 227$$

$$39x = 195$$

$$x = 5$$

$$x + 6 = 11 \quad\left.\right\} \quad \textit{Answers}$$

$$24x + 2 = 122$$

Check:

$$10(5) + 5(11) + 24(5) + 2 = 227$$

$$50 + 55 + 120 + 2 = 227$$

$$227 = 227$$

8. Let $\quad x = $ cost in cents of Tylenol
$x + 125 = $ cost in cents of aspirin
$2(x + x + 125) = $ cost in cents of cologne

Equation:

$$x + x + 125 + 2(2x + 125) = 2475$$
$$2x + 125 + 4x + 250 = 2475$$
$$6x + 375 = 2475$$
$$6x = 2100$$
$$x = 350$$
$$x + 125 = 475 \quad\left.\right\} \text{ } Answers$$
$$2(2x + 125) = 1650$$

Check:

$$3.50 + 4.75 + 16.50 = 24.75$$
$$24.75 = 24.75$$

9. Let
$$x = \text{number of 35-cent candy bars}$$
$$3x = \text{number of 10-cent mints}$$
$$3x + 1 = \text{number of 24-cent packages of gum}$$

Value in Cents

$$35x = \text{cents for candy bar}$$
$$10(3x) = \text{cents for mints}$$
$$24(3x + 1) = \text{cents for gum}$$

Equation:

$$35x + 10(3x) + 24(3x + 1) = 572$$
$$35x + 30x + 72x + 24 = 572$$
$$137x + 24 = 572$$
$$137x = 548$$
$$x = 4$$
$$3x = 12 \quad\left.\right\} \text{ } Answers$$
$$3x + 1 = 13$$

Check:

$$4(35) + 12(10) + 13(24) = 572$$

$$140 + 120 + 312 = 572$$

$$572 = 572$$

10. Let $\quad x$ = number of loaves of bread at \$2.39 each
$\quad\quad 2x$ = number of cartons of milk at \$1.95 each
$\quad x + 1$ = number of packages of cereal at \$3.00 each
$\quad x + 1$ = number of pounds of meat at \$5.39 each

Cost in Cents

$$239x = \text{cost of bread}$$

$$195(2x) = \text{cost of milk}$$

$$300(x + 1) = \text{cost of cereal}$$

$$539(x + 2) = \text{cost of meat}$$

Equation:

$$239x + 195(2x) + 300(x + 1) + 539(x + 1) = 5000 - 1225$$

$$239x + 390x + 300x + 300 + 539x + 539 = 3775$$

$$1468x + 839 = 3775$$

$$1468x = 2936$$

$$x = 2$$

$$2x = 4$$

$$x + 1 = 3$$

$$x + 1 = 3$$

$\left.\begin{array}{c} \\ \\ \\ \\ \end{array}\right\}$ *Answers*

Check:

$$239(2) + 195(4) + 300(3) + 539(3) = 5000 - 1225$$

$$478 + 780 + 900 + 1617 = 3775$$

$$3775 = 3775$$

Chapter 5

Age

Age problems usually follow a certain pattern. That is, they usually have certain basic facts. Frequently, they refer to ages at different points in time.

For example, Ted's father is twice as old as Ted. This means that their ages *now* could be represented as

$$x = \text{Ted's age now}$$

$$2x = \text{father's age now}$$

If the problem then states that 10 years ago the father was three times as old, you have to represent their ages 10 years ago. To show their ages 10 years ago, you don't need any other information. All you do is subtract 10 from their ages now. (If you know your age now, you subtract 10 to find your age 10 years ago.)

$$x - 10 = \text{Ted's age 10 years ago}$$

$$2x - 10 = \text{father's age 10 years ago}$$

Now you can use the fact that the father was *then* three times as old as Ted to set up the equation. The father's age 10 years ago was three times Ted's age 10 years ago.

$$2x - 10 = 3(x - 10)$$

This procedure of representing ages now, moving them forward or backward, and then setting up the equation is often used in age problems.

EXAMPLE 1

Mary's father is four times as old as Mary. Five years ago he was seven times as old. How old is each now?

Solution

The first fact is that Mary's father is four times as old as Mary.

Let x = Mary's age now   (smaller number)
    $4x$ = father's age now

The second fact is that 5 years ago . . . (stop!). The problem tells you about their ages at *another time*. Five years ago your age would be 5 less than your age now. So

$$x - 5 = \text{Mary's age 5 years ago}$$

$$4x - 5 = \text{father's age 5 years ago}$$

Now read on. Five years ago he was *seven times as old* (as she was).

$$4x - 5 = 7(x - 5)$$

$$4x - 5 = 7x - 35$$

$$-3x = -30$$

$$x = 10 \left.\vphantom{\begin{matrix}a\\b\end{matrix}}\right\} \quad Answers$$

$$4x = 40$$

Check:

$$x - 5 = 5$$

$$4x - 5 = 35$$

and

$$7 \times 5 = 35$$

EXAMPLE 2

Abigail is 6 years older than Jonathan. Six years ago she was twice as old as him. How old is each now?

92

Let x = Jonathan's age now   (smaller number)
   $x + 6$ = Abigail's age now
Then

$$x - 6 = \text{Jonathan's age 6 years ago}$$

$$(x + 6) - 6 = \text{Abigail's age 6 years ago}$$

Six years ago she was twice as old as him.

Equation:

$$(x + 6) - 6 = 2(x - 6)$$

$$x = 2x - 12$$

$$-x = -12$$

$$\left.\begin{array}{r} x = 12 \\ x + 6 = 18 \end{array}\right\} \quad Answers$$

Note that in this type of problem, as soon as you come to the statements about a time other than now, stop and move the ages *now* forward or backward. For "three years ago," you subtract 3 from ages now. For "five years from now," you add 5 to ages now. This is like saying that in 5 years you will be 5 years older. Be careful to use *only* the number of years you add or subtract and don't use any other information until the equation. The equation is usually based on the relationship in the future or past *after* you move the ages ahead or back.

EXAMPLE 3

Wolfgang's father is 26 years older than Wolfgang. In 10 years, the sum of their ages will be 80. What are their present ages?

Solution

Let x = Wolfgang's age now
   $x + 26$ = father's age now

Then $x + 10 =$ Wolfgang's age in 10 years
     $(x + 26) + 10 =$ father's age in 10 years

Equation:

$$x + 10 + (x + 26) + 10 = 80$$

$$2x + 46 = 80$$

$$2x = 34$$

$$\left. \begin{array}{l} x = 17 \\ x + 26 = 43 \end{array} \right\} \text{ \textit{Answers}}$$

Check: In 10 years Wolfgang will be 27 and his father will be 53.

$$53 + 27 = 80$$

EXAMPLE 4

Mary is twice as old as Helen. If 8 is subtracted from Helen's age and 4 is added to Mary's age, Mary will then be four times as old as Helen. Find their ages.
Note: This is a simple number relationship problem.

Solution

Let $x =$ Helen's age now
    $2x =$ Mary's age now
Then

$$x - 8 = 8 \text{ subtracted from Helen's age}$$

$$2x + 4 = 4 \text{ added to Mary's age}$$

Equation:

$$2x + 4 = 4(x - 8)$$

$$-2x = -36$$

94

$$\left. \begin{array}{r} x = 18 \\ 2x = 36 \end{array} \right\} \textit{Answers}$$

Check:

$$2(18) + 4 = 4(18 - 8)$$
$$36 + 4 = 4(10)$$
$$40 = 40$$

Supplementary Age Problems

1. A man is four times as old as his son. In 3 years, the father will be three times as old as his son. How old is each now?
2. Abigail is 8 years older than Cynthia. Twenty years ago Abigail was three times as old as Cynthia. How old is each now?
3. Seymour is twice as old as Cassandra. If 16 is added to Cassandra's age and 16 is subtracted from Seymour's age, their ages will then be equal. What are their present ages?
4. In 4 years Cranston's age will be the same as Terrill's age now. In 2 years, Terrill will be twice as old as Cranston. Find their ages now.
5. Mrs. Smythe is twice as old as her daughter Samantha. Ten years ago the sum of their ages was 46 years. How old is Mrs. Smythe?
6. A Roman statue is three times as old as a Florentine statue. One hundred years from now the Roman statue will be twice as old. How old is the Roman statue?
7. Sheri's age in 20 years will be the same as Terry's age is now. Ten years from now, Terry's age will be twice Sheri's. What are their present ages?
8. Bettina's age is three times Melvina's. If 20 is added to Melvina's age and 20 is subtracted from Bettina's, their ages will be equal. How old is each now?

Solutions to Supplementary Age Problems

1. Let $x =$ son's age now
$4x =$ father's age now

Then

$$x + 3 = \text{son's age in 3 years}$$

$$4x + 3 = \text{father's age in 3 years}$$

In 3 years, the father will be three times as old as his son.

Equation:

$$4x + 3 = 3(x + 3) = 3x + 9$$

$$\left.\begin{array}{r} x = 6 \\ 4x = 24 \end{array}\right\} \text{Answers}$$

2. Let $\quad x = \text{Cynthia's age now}$
$\qquad x + 8 = \text{Abigail's age now}$

Then

$$x - 20 = \text{Cynthia's age 20 years ago}$$

$$(x + 8) - 20 = \text{Abigail's age 20 years ago}$$

Equation:

$$(x + 8) - 20 = 3(x - 20)$$

$$x - 12 = 3x - 60$$

$$-2x = -48$$

$$\left.\begin{array}{r} x = 24 \\ x + 8 = 32 \end{array}\right\} \text{Answers}$$

3. Let $\ x = \text{Cassandra's age now}$
$\qquad 2x = \text{Seymour's age now}$

Then

$$x + 16 = 16 \text{ added to Cassandra's age}$$

$$2x - 16 = 16 \text{ subtracted from Seymour's age}$$

Equation:

$$2x - 16 = x + 16$$

$$\left.\begin{array}{r} x = 32 \\ 2x = 64 \end{array}\right\} \text{Answers}$$

4. Let x = Terrill's age now (and Cranston's age in 4 years)
$x - 4$ = Cranston's age now

Then

$$x + 2 = \text{Terrill's age in 2 years}$$

$$(x - 4) + 2 = \text{Cranston's age in 2 years}$$

Terrill will be twice as old as Cranston in 2 years.

Equation:

$$x + 2 = 2[(x - 4) + 2]$$
$$= 2x - 8 + 4$$
$$= 2x - 4$$
$$\left. \begin{array}{l} x = 6 \\ x - 4 = 2 \end{array} \right\} \text{Answers}$$

5. Let x = Samantha's age now
$2x$ = Mrs. Smythe's age now

Then

$$x - 10 = \text{Samantha's age 10 years ago}$$

$$2x - 10 = \text{Mrs. Smythe's age 10 years ago}$$

Equation:

$$x - 10 + 2x - 10 = 46$$
$$3x - 20 = 46$$
$$3x = 66$$
$$\left. \begin{array}{l} x = 22 \\ 2x = 44 \end{array} \right\} \text{Answers}$$

6. Let x = age of Florentine statue now
$3x$ = age of Roman statue now

Then

$$x + 100 = \text{age of Florentine statue in 100 years}$$

$$3x + 100 = \text{age of Roman statue in 100 years}$$

Equation:

$$3x + 100 = 2(x + 100) = 2x + 200$$
$$3x - 2x = 200 - 100$$
$$x = 100 \text{ years}$$
$$3x = 300 \text{ years}$$

$\left.\right\}$ *Answers*

7. Let $x = $ Terry's age now
    $x - 20 = $ Sheri's age now
Then

$$x + 10 = \text{Terry's age in 10 years}$$
$$(x - 20) + 10 = \text{Sheri's age in 10 years}$$

Equation:

$$x + 10 = 2[(x - 20) + 10]$$
$$= 2(x - 10)$$
$$= 2x - 20$$
$$-x = -30$$
$$x = 30$$
$$x - 20 = 10$$

$\left.\right\}$ *Answers*

8. Let $x = $ Melvina's age now
    $3x = $ Bettina's age now
Then

$$x + 20 = 20 \text{ added to Melvina's age}$$
$$3x - 20 = 20 \text{ subtracted from Bettina's age}$$

Equation:

$$x + 20 = 3x - 20$$
$$-2x = -40$$
$$x = 20$$
$$3x = 60$$

$\left.\right\}$ *Answers*

Chapter 6

Levers

Most children are familiar with teeter-totters or seesaws. They know that if big Sam sits on one end and little Jane sits on the other, Jane goes up in the air and Sam sinks to the ground. So Sam slides toward the center (called the balance point or *fulcrum*) until they are in balance.

This example illustrates the fact that in order to balance, unequal weights (or forces) must be different distances from the point of balance. The teeter-totter, board, or rod is called the *lever*. The principle involved is that the weight times its distance from the fulcrum on one side equals the weight times *its* distance from the fulcrum on the other side. If you let w_1 and d_1 represent one weight and distance, and w_2 and d_2 represent the others, you might diagram this relationship like this:

$$w_1 d_1 = w_2 d_2$$

If a 100-pound (lb) weight 4 feet (ft) from the fulcrum balances a 50-pound weight 8 feet from the fulcrum, then:

$$100 \text{ pounds} \times 4 \text{ feet} = 50 \text{ pounds} \times 8 \text{ feet}$$

$$400 = 400$$

In lever problems it is always wise to draw a sketch to show facts. If you know three of the four facts in the general equation $w_1d_1 = w_2d_2$, you can set up your own equation and solve for the fourth.

EXAMPLE I

Mary weighted 120 pounds and sat on one end of a teeter-totter 8 feet from the center. Jim weighted 160 pounds and sat on the other side. How far from the fulcrum (balance point) must Jim sit to balance Mary if the balance point is at the center of the board?

Solution

1. Read the problem and determine the unknown. The question asks how far from the fulcrum Jim sits, so

   Let x = Jim's distance from fulcrum in feet
2. **First fact.** Mary weighs 120 pounds and sits on one end of the teeter-totter.

3. **Second fact.** She sits 8 feet from the center.

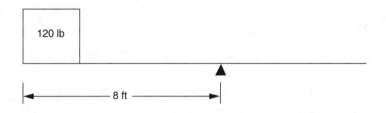

4. **Third fact.** Jim weighed 160 pounds and sat on the other side. We have already found that we would let x equal Jim's distance.

Equation:

$$w_1 d_1 = w_2 d_2$$

$$160x = 120(8) = 960$$

$$x = 6 \quad \textit{Answer}$$

EXAMPLE 2

Tim was digging up his yard to plant a garden. He hit a large rock and got a crowbar to raise the rock to the surface. The crowbar was 6 feet long and the rock was later found to weigh 80 pounds. Tim used another rock as the fulcrum and exerted a force equal to 16 pounds on the end of the crowbar in order to raise the rock to a point of balance. How far was the fulcrum from the 80-pound rock?

Solution

Notice that we again use the principle that the total minus one part equals the second part.

Let x = distance in feet of 80-pound rock from fulcrum

Equation:

$$w_1 d_1 = w_2 d_2$$
$$16(6 - x) = 80x$$
$$96 - 16x = 80x$$
$$-96x = -96$$
$$x = 1 \quad \textit{Answer}$$

EXAMPLE 3

A weight of 60 pounds rests on the end of an 8-foot lever and is 3 feet from the fulcrum. What weight must be placed on the other end of the lever to balance the 60-pound weight?

Solution

If the lever is 8 feet long and one end is 3 feet from the fulcrum, the other end must be 5 feet from the fulcrum $(8 - 3 = 5)$.

Let x = unknown weight in pounds

Equation:

$$5x = 60(3) = 180$$

$$x = 36 \quad Answer$$

EXAMPLE 4

Four girls decided to use the same teeter-totter. Two of them weighing 75 pounds and 50 pounds, respectively, sat on opposite ends of the 12-foot board which had the balance point at the center. The third girl weighing 60 pounds got on the same side as the one weighing 50 pounds and sat 5 feet from the fulcrum. Where must the fourth girl sit so the teeter-totter is balanced if she weighs 40 pounds?

Solution

Here we have something new. There is more than one weight on the same side. When this happens, you must add the *products* on the same side or

$$w_1d_1 + w_2d_2 = w_3d_3 + w_4d_4$$

Each weight is multiplied by its distance from the fulcrum. Let's take it one piece at a time. The girls weighing 75 and 50 pounds sat on opposite ends of a 12-foot board.

The one weighing 60 pounds got on the same side as the one weighing 50 pounds and sat 5 feet from the fulcrum.

$$w_1d_1 + w_2d_2 = w_3d_3$$

$$50(6) + 60(5) = 75(6)$$

If this equation is true, they balance. But 300 plus 300 is not equal to 450, so the fourth girl must sit on the right side. Her distance will be equal to x because that is the one fact we do not know.

Let x = distance of 40-pound girl from fulcrum in feet

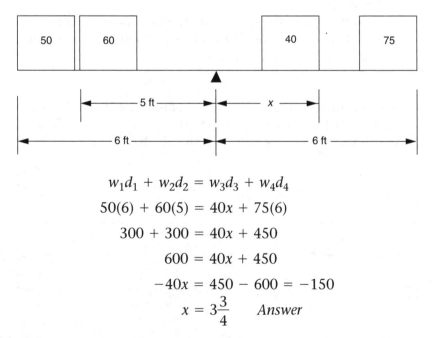

$$w_1d_1 + w_2d_2 = w_3d_3 + w_4d_4$$

$$50(6) + 60(5) = 40x + 75(6)$$

$$300 + 300 = 40x + 450$$

$$600 = 40x + 450$$

$$-40x = 450 - 600 = -150$$

$$x = 3\frac{3}{4} \qquad Answer$$

Supplementary Lever Problems

1. A 200-pound weight rests on one end of a lever, 12 feet from the fulcrum. What weight, resting on the opposite end and 3 feet from the fulcrum, would make a balance?

2. The boys take a 12-foot-long board and rest it on a large rock to make a teeter-totter. If the boys sit on opposite ends of the board and weigh 50 pounds and 70 pounds, respectively, how far from the fulcrum will the 70-pound boy be if they balance?

3. A window in Mr. Jones's house is stuck. He takes an 8-inch screw driver to pry open the window. If the screwdriver rests on the sill (fulcrum) 3 inches from the window and Mr. Jones has to exert a force of 10 pounds on the other end to pry open the window, how much force was the window exerting?

4. A 35-pound weight is 2 feet from the fulcrum, and a 75-pound weight on the same side is 10 feet from the fulcrum. If a weight on the other end 6 feet from the fulcrum balances the first two, how much does it weigh?

5. A lever 10 feet long has a 100-pound weight on one end and a 150-pound weight on the other. If the fulcrum is in the center, in what location must an 80-pound weight be placed so that the lever will balance?

6. Shelly and Karie go out to play. Shelly, who weighs 90 pounds, sits on one end of a 14-foot teeter-totter. Its balance point is at the center of the board. Karie, who weighs 120 pounds, climbs on the other end and slides toward the center until they balance. What is Karie's distance from her end of the teeter-totter when they balance?

Solutions to
Supplementary Lever Problems

1. Let x = weight in pounds to make a balance

$$3x = 12(200) = 2400$$

$$x = 800 \quad \textit{Answer}$$

2. Let x = distance in feet of 70-pound boy from fulcrum

$$50(12 - x) = 70x$$

$$600 - 50x = 70x$$

$$-120x = -600$$

$$x = 5 \quad \textit{Answer}$$

3. Let x = force the window exerts in pounds

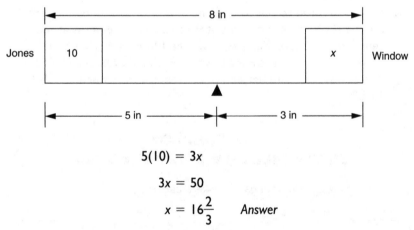

$$5(10) = 3x$$

$$3x = 50$$

$$x = 16\frac{2}{3} \quad \textit{Answer}$$

4. Let x = weight 6 feet from fulcrum in pounds

$$6x = 2(35) + 10(75)$$
$$= 70 + 750$$
$$= 820$$
$$x = 136\frac{2}{3} \quad \textit{Answer}$$

5.

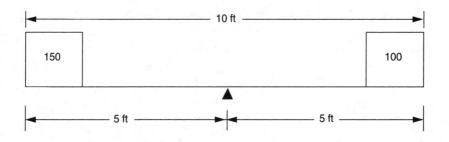

Find products:

$$5(150) = 750 \text{ on left}$$
$$5(100) = 500 \text{ on right}$$

Therefore the extra weight must be put on the right side to maintain balance.

Let x = distance of 80-pound weight from fulcrum in feet

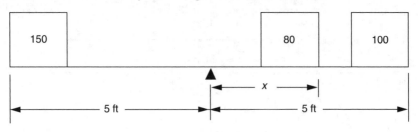

Equation:

$$5(150) = 80x + 5(100)$$
$$750 = 80x + 500$$
$$80x + 500 = 750$$
$$80x = 250$$
$$x = 3\frac{1}{8} \qquad Answer$$

6. Let x = Karie's distance from the center in feet

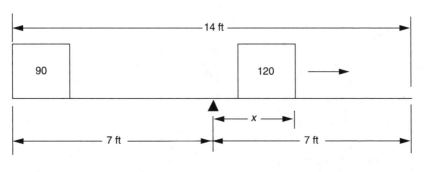

$$(90)(7) = 120x$$
$$120x = 630$$
$$x = 5\frac{1}{4}$$
$$7 - 5\frac{1}{4} = 1\frac{3}{4} \qquad Answer$$

The question this time asked for the distance of the 120-pound weight (Karie's weight) from the end. You must find the distance from the fulcrum first and subtract this distance from the total of 7 feet to find the distance from the end of the teeter-totter.

Chapter 7

Finance

There are many types of problems involving money. They should be easy to understand because their principles are familiar to everyone. Let's start with some investment problems. You should know that investments involve three basics: the *amount of money invested* (the principal), the *rate of interest paid* (the rate of return in percent), and the actual yearly *interest in dollars* (often referred to as the return on the investment). *Interest, income,* and *return* all refer to the same thing for these problems and any may be used in the examples and problems. (Assume simple interest compounded yearly.)

EXAMPLE I

Mr. McGregor invested $10,000. Part of it he put in the bank at 5 percent interest. The remainder he put in bonds which pay a 9 percent yearly return. How much did he invest in each vehicle if his yearly income from the two investments was $660?

Solution

Let x = amount in dollars invested at 5 percent
$10,000 - x$ = amount in dollars invested at 9 percent

Now that you have represented the unknowns, using x and total $- x$, and labeled them with their *percent* return, you have to convert the information into *income*. Here is the way you find income (interest): principal times rate equals interest.

$$P \times r = I$$

109

If you put $1000 in the bank at 5 percent, your interest would be 1000(0.05); remember 5 percent = 0.05. Setting up our equation,

$$0.05x = \text{interest earned from bank investment}$$

$$0.09(10,000 - x) = \text{interest earned from bond investment}$$

Equation: The interest earned from the bank plus the interest earned from the bond equals the total interest earned.

$$0.05x + 0.09(10,000 - x) = 660$$

In solving an equation containing parentheses and decimals, first clear parentheses and then clear decimals. This will help avoid careless errors in decimals. Multiplying 0.09 times 10,000 and 0.09 times x you get

$$0.05x + 900 - 0.09x = 600$$

Multipy both sides of the equation by 100 to clear the decimals.

$$5x + 90,000 - 9x = 66,000$$

Group like terms.

$$-4x = -24,000$$

Divide both sides by -4.

$$\left. \begin{array}{r} x = 6000 \\ 10,000 - x = 4000 \end{array} \right\} \; Answers$$

Check:

$$0.05(6000) + 0.09(4000) = 660$$

$$300 + 360 = 660$$

$$660 = 660$$

EXAMPLE 2

Mr. Gold invested $50,000, part at 6 percent and part at 8 percent. The annual interest on the 6 percent investment was $480 more than that from the 8 percent investment. How much was invested at each rate?

Solution

Let

x = amount in dollars invested at 6 percent

$50,000 - x$ = amount in dollars invested at 8 percent

Then

$$0.06x = \text{income on first investment}$$

$$0.08(50,000 - x) = \text{income on second investment}$$

The interest on the 6 percent investment was $480 more than the interest on the 8 percent investment.

Equation: The first income equals the second income plus 480.

$$0.06x = 0.08(50,000 - x) + 480$$

$$= 4000 - 0.08x + 480$$

$$= 4480 - 0.08x$$

$$6x = 448,000 - 8x$$

$$14x = 448,000$$

$$\left.\begin{array}{l} x = 32,000 \\ 50,000 - x = 18,000 \end{array}\right\} \textit{Answers}$$

Check:

$$0.06(32,000) = 0.08(18,000) + 480$$

$$1920 = 1440 + 480 = 1920$$

EXAMPLE 3

A store advertised dresses on sale at 20 percent off. The sale price was $76. What was the original price of the dress?

Solution

Let x = original price in dollars

If the sale price is 20 percent *off* the original price, the sale price is 80 percent of the original price.

Equation:

$$0.80x = 76$$

$$80x = 7600$$

$$x = 95 \qquad \textit{Answer}$$

EXAMPLE 4

Tickets to the school play sold at $4 each for adults and $1.50 each for children. If there were four times as many adult tickets sold as children's tickets, and the total receipts were $3500, how many children's tickets were sold?

Solution

Let x = number of children's tickets at $1.50 each

$\quad 4x$ = number of adult tickets at $4 each

Then

$\qquad 1.50x$ = total amount of money received from children's tickets in dollars

$\qquad 4(4x)$ = total amount of money received from adult tickets in dollars

Equation:

$$1.50x + 4(4x) = 3500$$

$$1.50x + 16x = 3500$$

$$15x + 160x = 35{,}000$$

$$175x = 35,000$$

$$x = 200 \qquad \textit{Answer}$$

EXAMPLE 5

Mr. Herkowitz owns a jewelry store. He marks up all merchandise 50 percent of cost. If he sells a diamond ring for $1500, what did he pay the wholesaler for it?

Solution

Let x = amount in dollars he paid wholesaler
$0.50x$ = his markup

Cost plus markup equals selling price.

Equation:

$$x + 0.50x = 1500$$

Multiply by 10 to clear decimals.

$$10x + 5x = 15,000$$

$$15x = 15,000$$

$$x = 1000 \qquad \textit{Answer}$$

EXAMPLE 6

What amount of money invested at $8\frac{1}{4}$ percent yields a $2475 return per year?

Solution

Let x = amount of money invested in dollars
$0.0825x$ = interest of $8\frac{1}{4}$ percent per year

Equation:

$$0.0825x = 2475$$

$$x = 30,000$$

113

EXAMPLE 7

Large supermarkets often price miscellaneous merchandise at a price they know will sell fast and determine what they can pay for it by this selling price. The Piggley Market bought men's T-shirts to sell at $10 each. If they allow 40 percent of the *selling price* for expenses and profit, what will they be willing to pay for the shirts?

Solution

Cost plus the markup equals the selling price.

Let x = cost in dollars they will pay for each shirt
$0.40(10)$ = markup (40 percent of selling price)

Equation:

$$x + 0.40(10) = 10$$

$$x + 4 = 10$$

$$x = 6 \qquad Answer$$

Check:

$$6 + 0.40(10) = 10$$

$$6 + 4 = 10$$

$$10 = 10$$

Alternate Solution

Let x = cost in dollars of each shirt
Cost + markup = selling price

If the markup is 40 percent of the selling price, then the cost is 60 percent of the selling price.

Cost (60 percent) + markup (40 percent) = selling price (100 percent)

Cost = 60 percent of the selling price

Equation:

$$x = 0.60(10)$$

$$x = 6 \quad Answer$$

Supplementary Finance Problems

1. Mr. Phinance invested $50,000. Part of it he put in a gold mine stock from which he hoped to receive a 20 percent return per year. The rest he invested in a bank stock which was paying 6 percent per year. If he received $400 more the first year from the bank stock than from the mining stock, how much did he invest in each stock?

2. Mr. Jordan wished to invest a sum of money so that the interest each year would pay his son's college expenses. If the money was invested at 8 percent and the college expenses were $10,000 per year, how much should Mr. Jordan invest?

3. The Worn & Torn Store had a sale on T-shirts at which all shirts were sold at 15 percent off the original price. Casey bought a shirt for $7.65. What was the original selling price of the shirt? (Assume no taxes.)

4. Tom inherited two different stocks whose yearly income was $2100. The total appraised value of the stocks was $40,000; one was paying 4 percent and one 6 percent per year. What was the value of each stock?

5. A men's store bought 500 suits, some at $125 each and the rest at $200 each. If the total cost of the suits was $77,500, how many suits were purchased at each price?

6. When Mrs. Middleclass sold her house recently, she received $210,000 for it. This was 40 percent more than she paid for it 10 years ago. What was the original purchase price?

7. Mrs. Lucky inherited $20,000 which she invested in stocks and bonds. The stocks returned 6 percent and the bonds 8 percent. If the return on the bonds was $80 less than the return on the stocks, how much did she invest in each?

8. The total of two investments is $25,000. One amount is invested at 7 percent and one at 9 percent. The annual interest from the 7 percent investment is $470 more than from the 9 percent investment. How much is invested at each rate?

9. A taxpayer's state and federal income taxes plus an inheritance tax totaled $14,270. His California state income tax was $5780 less than his federal tax. His inheritance tax was $2750. How much did he pay in state and federal taxes?

10. Mrs. Scott had saved $6000 which she wished to invest. She put part of it in a term Certificate of Deposit (CD) at 8 percent and part in a regular savings account at $5\frac{1}{2}$ percent. How much was invested in each account if her total yearly income amounted to $425?

11. Dr. Wellness had $10,000 invested at 5 percent. How many dollars more would he have to invest at 8 percent so that his total interest per year would equal 7 percent of the two investments?

Solutions to
Supplementary Finance Problems

1. Let

$$x = \text{amount in dollars invested at 20 percent}$$
$$50,000 - x = \text{amount in dollars invested at 6 percent}$$
$$\text{(total minus } x\text{)}$$
$$0.20x = \text{interest on mining stock (smaller income)}$$
$$0.06(50,000 - x) = \text{interest on bank stock (larger income)}$$

Equation: Income from the bank stock is interest on the mining stock plus $400.

$$0.06(50,000 - x) = 0.20x + 400$$

$$3000 - 0.06x = 0.20x + 400$$

$$300,000 - 6x = 20x + 40,000$$

$$-26x = -260,000$$

$$\left. \begin{array}{c} x = 10,000 \\ 50,000 - x = 40,000 \end{array} \right\} \text{ Answers}$$

Check:

$$0.06(40,000) = 0.20(10,000) + 400$$

$$2400 = 2000 + 400 = 2400$$

2. Let x = amount in dollars Michael should invest
$0.08x$ = interest per year (assume simple interest)

Equation:

$$0.08x = 10,000$$
$$8x = 1,000,000$$
$$x = 125,000 \quad \textit{Answer}$$

Check:

$$0.08(125,000) = 10,000$$
$$10,000 = 10,000$$

3. Let x = original selling price in dollars
$0.15x$ = discount
The original price minus the discount equals the sale price.

Equation:

$$x - 0.15x = 7.65$$
$$0.85x = 7.65$$
$$85x = 765$$
$$x = 9 \quad \textit{Answer}$$

Check:

$$9 - 0.15(9) = 7.65$$
$$9 - 1.35 = 7.65$$
$$7.65 = 7.65$$

4. Let x = value in dollars of stock paying 4 percent
$40,000 - x$ = value in dollars of stock paying 6 percent
$0.04x$ = interest on stock paying 4 percent
$0.06(40,000 - x)$ = interest on stock paying 6 percent

Equation:

$$0.04x + 0.06(40,000 - x) = 2100$$

$$0.04x + 2400 - 0.06x = 2100$$

$$4x + 240,000 - 6x = 210,000$$

$$-2x = -30,000$$

$$\left. \begin{array}{l} x = 15,000 \\ 40,000 - x = 25,000 \end{array} \right\} \; Answers$$

Check:

$$0.04(15,000) + 0.06(25,000) = 2100$$

$$600 + 1500 = 2100$$

$$2100 = 2100$$

5. Let x = number of suits purchased at $125 each
$500 - x$ = number of suits purchased at $200 each

The total value equals the price times the number at $125 plus the price times the number at $200.

Equation:

$$125x + 200(500 - x) = 77,500$$

$$125x + 100,000 - 200x = 77,500$$

$$-75x = 22,500$$

$$\left. \begin{array}{ll} x = 300 & \text{(suits at \$125 each)} \\ 500 - x = 200 & \text{(suits at \$200 each)} \end{array} \right\} \; Answers$$

Check:

$$125(300) + 200(200) = 77,500$$

$$37,500 + 40,000 = 77,500$$

$$77,500 = 77,500$$

6. Let $x =$ original purchase price in dollars
$0.40x =$ increase in value in dollars

Equation:

$$x + 0.40x = 210,000$$

$$1.40x = 210,000$$

$$14x = 2,100,000$$

$$x = 150,000 \quad \text{(original purchase price)} \qquad \textit{Answer}$$

Check:

$$150,000 = 0.4(150,000) = 210,000$$

$$150,000 + 60,000 = 210,000$$

$$210,000 = 210,000$$

7. Let $\quad x =$ amount in dollars invested in stock at 6 percent
$20,000 - x =$ amount in dollars invested in bonds at 8 percent
$0.06x =$ interest on stocks
$0.08(20,000 - x) =$ interest on bonds

Equation: Interest on stocks equals interest on bonds plus $80.

$$0.06x = 0.08(20,000 - x) + 80$$

$$= 1600 - 0.08x + 80$$

$$6x = 160,000 - 8x + 8000$$

$$14x = 168,000$$

$$x = 12,000 \quad \text{(stocks)}$$
$$20,000 - x = 8000 \quad \text{(bonds)} \qquad \textit{Answers}$$

Check:

$$0.06(12,000) = 0.08(8000) + 80$$

$$720 = 640 + 80 = 720$$

8. Let $\qquad x =$ amount in dollars invested at 7 percent
$\quad 25{,}000 - x =$ amount in dollars invested at 9 percent

Equation: The amount at 7 percent equals the amount at 9 percent plus $470.

$$0.07x = 0.09(25{,}000 - x) + 470$$

$$0.07x = 2250 - 0.09x + 470$$

$$7x = 225{,}000 - 9x = 47{,}000$$

$$16x = 272{,}000$$

$$x = 17{,}000 \quad \text{(amount invested at 7 percent)}$$

$$25{,}000 - x = 8000 \quad \text{(amount invested at 9 percent)} \qquad \left.\right\}\ \textit{Answers}$$

Check:

$$0.07(17{,}000) = 0.09(8000) + 470$$

$$1190 = 720 + 470$$

$$1190 = 1190$$

9. Let $\qquad x =$ amount in dollars of state tax
$\quad x + 5780 =$ amount in dollars of federal tax

Equation:

$$x + x + 5780 + 2750 = 14{,}270$$

$$2x + 8530 = 14{,}270$$

$$2x = 5740$$

$$x = 2870 \quad \text{(state tax)}$$

$$x + 5780 = 8650 \quad \text{(federal tax)} \qquad \left.\right\}\ \textit{Answers}$$

Check:

$$2870 + 2870 + 5780 + 2750 = 14{,}270$$

$$14{,}270 = 14{,}270$$

120

10. Let x = amount in dollars invested at 8 percent

$6000 - x$ = amount in dollars invested at $5\frac{1}{2}$ percent

$0.08x$ = interest on 8 percent investment

$0.055(6000 - x)$ = interest on $5\frac{1}{2}$ percent investment

Equation: Total interest (income) was $425.

$$0.08x + 0.055(6000 - x) = 420$$

$$0.08x + 330 - 0.055x = 425$$

Multiply by 1000 to clear decimals.

$$80x + 330,000 - 55x = 425,000$$

$$25x = 95,000$$

$$x = 3800 \Big\}$$
$$6000 - x = 2200 \Big\} \quad Answers$$

Check:

$$0.08(3800) + 0.055(2200) = 425$$

$$304 + 121 = 425$$

$$425 = 425$$

11. Let x = amount in dollars at 8 percent

$10,000$ = total amount invested at 5 percent

$x + 10,000$ = total amount invested at 7 percent

$0.08x$ = interest on amount at 8 percent

$0.05(10,000)$ = interest on amount at 5 percent

$0.07(x + 10,000)$ = interest on entire investment at 7 percent

Equation: Total interest per year equals 7 percent of the entire investment.

$$0.08x + 0.05(10,000) = 0.07(x + 10,000)$$

$$0.08x + 500 = 0.07x + 700$$

Multiply both sides by 100.

$$8x + 50{,}000 = 7x + 70{,}000$$

$$x = 20{,}000$$

Check:

$$0.08(20{,}000) + 0.05(10{,}000) = 0.07(30{,}000)$$

$$1600 + 500 = 2100$$

$$2100 = 2100$$

Chapter 8

Work

Now let's take a look at problems involving people or machines doing a job. These are called work problems. Laborers may be digging a ditch, secretaries typing papers, pipes filling a tank. We shall start with basic problems. For example, Mr. Jones can build a garage in 3 days, and Mr. Smith can build the same garage in 5 days. How long would it take them to build the garage working together?

Before we solve this problem, there are certain facts about work problems to remember.

1. The jobs in each problem are either the same or equivalent jobs. In the problem above, Smith builds a garage exactly the same as the one Jones builds.
2. In solving a work problem, you need to work with the same units of measure, such as minutes, hours, or days, within each problem. You can't mix minutes and hours in the same equation, for example.
3. You always need to find the *fractional part* of the job which would be done in one unit of time, such as 1 minute or 1 hour. If a person can do a complete job in 3 days, he can do one-third of it in 1 day.
4. The fractional part of the job one person can do in 1 day plus the fractional part another person can do in 1 day equals the fractional part of the job the two can do together in 1 day. If Mr. Jones builds one-third of the garage in 1 day and Mr. Smith builds one-fifth of it in 1 day, together they can build one-third plus one-fifth of the garage in 1 day.

We can make a diagram to help solve this type of problem also. Suppose we wanted to know how many days it would take them to build the garage together.

Let x = number of days to build the garage together

	Total time in days	Fractional part of job done in 1 day
Jones	3	$\dfrac{1}{3}$
Smith	5	$\dfrac{1}{5}$
Together	x	$\dfrac{1}{x}$

If you add the fractional parts of the job the two can do separately in 1 day, the answer will be equal to the fractional part of the job the two can do together in 1 day.

$$\frac{1}{3} + \frac{1}{5} = \frac{1}{x}$$

Multiply by the LCD, $15x$, to clear fractions.

$$5x + 3x = 15$$

$$8x = 15$$

$$x = 1\frac{7}{8} \quad \textit{Answer}$$

EXAMPLE 1

Bob can dig a ditch in 4 hours. John can dig the same ditch in 3 hours. How long would it take them to dig it together?

Solution

Let x = number of hours to dig the ditch together

	Total time in hours	Fractional part of job done in 1 hour
Bob	4	$\frac{1}{4}$
John	3	$\frac{1}{3}$
Together	x	$\frac{1}{x}$

If Bob takes 4 hours to dig the ditch, he can dig one-fourth of it in 1 hour. John can dig one-third of it in 1 hour. If it takes them x hours to dig it together, they can dig $\frac{1}{x}$ part of it in 1 hour together. The total of the fractional part each can dig or one-third plus one-fourth equals the fractional part they can dig together in 1 hour or $\frac{1}{x}$.

Equation:

$$\frac{1}{3} + \frac{1}{4} = \frac{1}{x}$$

Multiply by the LCD, $12x$, to clear fractions.

$$4x + 3x = 12$$

$$7x = 12$$

$$x = 1\frac{5}{7} \quad \textit{Answer}$$

EXAMPLE 2

If John can type a paper in 5 hours and together he and Jim can type it in 2 hours, how long would it take Jim to type the same paper alone?

Solution

Let x = number of hours for Jim to type alone

125

	Total time in hours	Fractional part of job done in 1 hour
John	5	$\dfrac{1}{5}$
Jim	x	$\dfrac{1}{x}$
Together	2	$\dfrac{1}{2}$

Equation:

$$\frac{1}{5} + \frac{1}{x} = \frac{1}{2}$$
$$2x + 10 = 5x$$
$$-3x = -10$$
$$x = 3\frac{1}{3} \qquad Answer$$

From these basic problems we can go to more complicated ones, but we will always work with fractional parts of a job done in 1 day, 1 hour, etc. The complete job is equal to the sum of the fractional parts or is equal to 1. If A does one-quarter, B does one-quarter, and C does one-half, the whole job is done, and $\dfrac{1}{4} + \dfrac{1}{4} + \dfrac{1}{2} = 1$.

EXAMPLE 3

Two women paint a barn. Barbara can paint it alone in 5 days, Sara in 8 days. They start to paint it together, but after 2 days Sara gets bored and Barbara finishes alone. How long does it take her to finish?

Solution

All work problems need diagrams. In the first column, show the total time for each person to do the job. Then in the next column show the fractional part of the job each person can do in *one* unit of time.

Let x = number of days for Barbara to finish barn

	Total time in days	Fractional part of job done in 1 day
Barbara	5	$\dfrac{1}{5}$
Sara	8	$\dfrac{1}{8}$

Next, for the third column multiply the time each person actually works by the column marked "Fractional part of job done in 1 day." The resulting fraction in the third column is the part of the job each person does in the time worked.

	Total time in days	Fractional part of job done in 1 day	Fractional part each does in 2 days	Fractional part Barbara does in x days
Barbara	5	$\dfrac{1}{5}$	$\dfrac{2}{5}$	$\dfrac{x}{5}$
Sara	8	$\dfrac{1}{8}$	$\dfrac{2}{8}$	—

Barbara does $\dfrac{2}{5}$, Sara $\dfrac{2}{8}$, and Barbara $\dfrac{x}{5}$ of the job. Add each part and you get one complete job, equal to 1.

Equation:

$$\frac{2}{5} + \frac{2}{8} + \frac{x}{5} = 1$$

Multiply by the LCD, 40, to clear fractions.

$$16 + 10 + 8x = 40$$

$$8x + 26 = 40$$

$$8x = 14$$

$$x = 1\frac{3}{4} \quad \textit{Answer}$$

EXAMPLE 4

A swimming pool can be filled by an inlet pipe in 10 hours and emptied by an outlet pipe in 12 hours. One day the pool is empty and the owner opens the inlet pipe to fill the pool. But he forgets to close the outlet. With both pipes open, how long will it take to fill the pool?

Solution

This time we have two pipes whose flows work against each other. The filling takes place faster than the emptying. So in any given time, the part of the job done is equal to the amount filled minus the amount emptied.

Let x = number of hours to fill pool with both pipes open

	Total time in hours	Fractional part filled or emptied in 1 hour	Fractional part each pipe does in x hours
Inlet pipe	10	$\dfrac{1}{10}$	$\dfrac{x}{10}$
Outlet pipe	12	$\dfrac{1}{12}$	$\dfrac{x}{12}$

Equation: Amount filled minus amount emptied equals whole job.

$$\frac{x}{10} - \frac{x}{12} = 1$$

Multiply by the LCD, 60, to clear fractions.

$$6x - 5x = 60$$

$$x = 60 \qquad \textit{Answer}$$

EXAMPLE 5

Mary, Sue, and Bill work at a motel. If each worked alone, it would take Mary 10 hours, Sue 8 hours, and Bill 12 hours to clean the whole motel. One day Mary came to work early and she had cleaned for 2 hours when Sue and Bill arrived and all three finished the job. How long did they take to finish?

Solution

Let x = number of hours they took to finish

	Total time in hours	Fractional part of job done in 1 hour	Fractional part Mary works in 2 hours	Fractional part each works in x hours
Mary	10	$\dfrac{1}{10}$	$\dfrac{2}{10}$	$\dfrac{x}{10}$
Sara	8	$\dfrac{1}{8}$	—	$\dfrac{x}{8}$
Bill	12	$\dfrac{1}{12}$	—	$\dfrac{x}{12}$

Equation: The fractional parts of the job each does added together equal the total job or 1.

$$\frac{2}{10} + \frac{x}{10} + \frac{x}{8} + \frac{x}{12} = 1$$

Multiply by the LCD, 120.

$$24 + 12x + 15x + 10x = 120$$

$$37x = 96$$

$$x = 2\frac{22}{37} \qquad Answer$$

Supplementary Work Problems

1. Tom, Dick, and Harry decided to fence a vacant lot adjoining their properties. If it would take Tom 4 days to build the fence, Dick 3 days, and Harry 6 days, how long would it take them working together?

2. John and Charles are linemen. John can string 10 miles of line in 3 days, while together they can string it in 1 day. How long would it take Charles alone to string the same line?

3. A firm advertises for workers to address envelopes. Priscilla says she will work 100 hours. Herb will work for 80 hours. If each can address 10,000 envelopes in the time they work, how long would it take them to address 10,000 envelopes if they work together?

4. Mrs. Hensen cataloged a group of incoming library books in 8 hours. Mr. Taylor can catalog the same number of books in 4 hours. Mrs. Phillips can catalog them in 6 hours. How long would it take if they all worked together?

5. One machine can complete a job in 10 minutes. If the same job is done by this machine and an older machine working together, the job can be completed in 6 minutes. How long would it take the older machine to do the job alone?

6. A pond is being drained by a pump. After 3 hours, the pond is half empty. A second pump is put into operation, and together the two pumps finish emptying the pond in half an hour. How long would it take the second pump to drain the pond if it had to do the same job alone?

7. Jim and his dad are bricklayers. Jim can lay bricks for a fireplace and chimney in 5 days. With his father's help he can build it in 2 days. How long would it take his father to build it alone?

8. Jones can paint a car in 8 hours. Smith can paint the same car in 6 hours. They start to paint the car together. After 2 hours, Jones leaves for lunch and Smith finishes painting the car alone. How long does it take Smith to finish?

9. Bill, Bob, and Barry are hired to paint signs. In 8 hours Bill can paint 1 sign, Bob can paint 2 signs, and Barry can paint $1\frac{1}{3}$ signs. They all come to work the first day, but Barry doesn't like the job and quits after 3 hours. Bob works half an hour longer than Barry and quits. How long will it take Bill to finish the two signs they were supposed to paint?

10. Keith has to grind the valves on his Mercedes. He estimates it will take him 8 hours. His friends Joe and Russ have done the same job in 10 and 12 hours, respectively. Keith had worked for 2 hours when Joe came along and helped him for 2 hours. Joe worked another half an hour alone and then Russ came along and together they finished the job. How long did it take them to finish?

11. Mr. Caleb hires Jay and Debbie to prune his vineyard. Jay can prune the vineyard in 50 hours, and Debbie can prune it in 40 hours. How many hours will it take them working together?

12. A high school hiking club held a car wash to raise money for equipment. Rudy, Cheryl, Tom, and Pat volunteered to help. Rudy

could wash a car in 10 minutes, Cheryl in 12, Tom in 8, and Pat in 15. They all started on the first car, but after 2 minutes another car came in and Pat and Tom went to work on it. One minute later Rudy quit to take care of another customer. How long did it take Cheryl to finish the first car alone?

Solutions to
Supplementary Work Problems

1. Let x = number of days to build fence together

	Total time in days	Fractional part of job done in 1 day
Tom	4	$\frac{1}{4}$
Dick	3	$\frac{1}{3}$
Harry	6	$\frac{1}{6}$
Together	x	$\frac{1}{x}$

Equation: The fractional parts each person does in 1 day added together equal the fractional part done together in 1 day.

$$\frac{1}{4} + \frac{1}{3} + \frac{1}{6} = \frac{1}{x}$$

$$3x + 4x + 2x = 12$$

$$9x = 12$$

$$x = 1\frac{1}{3} \qquad \textit{Answer}$$

2. Let x = number of days for Charles to string line alone

	Total time in days	Fractional part of job done in 1 day
John	3	$\dfrac{1}{3}$
Charles	x	$\dfrac{1}{x}$
Together	1	$\dfrac{1}{1}$

Equation:

$$\frac{1}{3} + \frac{1}{x} = \frac{1}{1}$$

$$x + 3 = 3x$$

$$-2x = -3$$

$$x = 1\frac{1}{2} \qquad \textit{Answer}$$

3. Let x = number of hours to address envelopes together

	Total time in hours	Fractional part of job done in 1 hour
Priscilla	100	$\dfrac{1}{100}$
Herb	80	$\dfrac{1}{80}$
Together	x	$\dfrac{1}{x}$

Equation:

$$\frac{1}{100} + \frac{1}{80} = \frac{1}{x}$$

Multiply by the LCD, 400x.

$$4x + 5x = 400$$

$$9x = 400$$

$$x = 44\frac{4}{9} \qquad \textit{Answer}$$

132

4. Let x = number of hours for all three working together to catalog books

	Total time in hours	Fractional part of job done in 1 hour
Hensen	8	$\dfrac{1}{8}$
Taylor	4	$\dfrac{1}{4}$
Phillips	6	$\dfrac{1}{6}$
Together	x	$\dfrac{1}{x}$

The total of the fractional parts of the job done by each in 1 hour equals the fractional part done by all of them working together.

Equation:

$$\frac{1}{8} + \frac{1}{4} + \frac{1}{6} = \frac{1}{x}$$

Multiply by the LCD, 24x, to clear fractions.

$$3x + 6x + 4x = 24$$

$$13x = 24$$

$$x = 1\frac{11}{13} \qquad \textit{Answer}$$

5. Let x = number of minutes for older machine to do job

	Total time in minutes	Fractional part of job done in 1 minute
First machine	10	$\dfrac{1}{10}$
Older machine	x	$\dfrac{1}{x}$
Together	6	$\dfrac{1}{6}$

Equation:

$$\frac{1}{10} + \frac{1}{x} = \frac{1}{6}$$

Multiply by the LCD, 30x, to clear fractions.

$$3x + 30 = 5x$$

$$-2x = -30$$

$$x = 15 \qquad \textit{Answer}$$

6. If the first pump drains half the pond in 3 hours, it could drain the pond completely in 6 hours. If the two pumps drain the last half of the pond in half an hour, together they could drain the whole pond in 1 hour.

Let x = number of hours it would take second pump to empty pond

	Total time in hours	Fractional part of job done in 1 hour
First pump	6	$\dfrac{1}{6}$
Second pump	x	$\dfrac{1}{x}$
Together	1	$\dfrac{1}{1}$

Equation:

$$\frac{1}{6} + \frac{1}{x} = \frac{1}{1}$$

Multiply by the LCD, 6x, to clear fractions.

$$x + 6 = 6x$$

$$-5x = -6$$

$$x = 1\frac{1}{5} \qquad \textit{Answer}$$

7. Let x = number of days it would take father alone

134

	Total time in days	Fractional part of job done in 1 day
Jim	5	$\dfrac{1}{5}$
Father	x	$\dfrac{1}{x}$
Together	2	$\dfrac{1}{2}$

Equation: The fractional part Jim can do plus the fractional part his father can do equal the fractional part they can do together.

$$\frac{1}{5} + \frac{1}{x} = \frac{1}{2}$$

Multiply by the LCD, $10x$, to clear fractions.

$$2x + 10 = 5x$$

$$-3x = -10$$

$$x = 3\frac{1}{3} \qquad \textit{Answer}$$

8. Let x = number of hours it takes Smith to finish

	Total time to do job in hours	Fractional part of job done in 1 hour	Fractional part each does in: 2 hours	x hours
Jones	8	$\dfrac{1}{8}$	$2\left(\dfrac{1}{8}\right)$	—
Smith	6	$\dfrac{1}{6}$	$2\left(\dfrac{1}{6}\right)$	$x\left(\dfrac{1}{6}\right)$

Jones and Smith each work for 2 hours. In that time each did two times the fractional part he could do in 1 hour or $2\left(\dfrac{1}{8}\right)$ and $2\left(\dfrac{1}{6}\right)$. Jones left and *only* Smith worked for x hours, so he did $x\left(\dfrac{1}{6}\right)$ of the job. When the job is completed, all the fractional parts of the job done add up to 1.

Equation:

$$2\left(\frac{1}{8}\right) + 2\left(\frac{1}{6}\right) + x\left(\frac{1}{6}\right) = 1$$

$$\frac{2}{8} + \frac{2}{6} + \frac{x}{6} = 1$$

Reduce fractions.

$$\frac{1}{4} + \frac{1}{3} + \frac{x}{6} = 1$$

Multiply by the LCD, 12.

$$3 + 4 + 2x = 12$$

$$2x + 7 = 12$$

$$2x = 5$$

$$x = 2\frac{1}{2} \quad \textit{Answer}$$

9. Let x = number of hours it takes Bill to finish two signs

The total job in this problem is painting two signs. So the total time for each person is the time it takes him to paint two signs. Bill can paint one sign in 8 hours or two signs in 16 hours. Bob can paint two signs in 8 hours. Barry can paint $1\frac{1}{3}$ signs in 8 hours.

$$8 \div 1\frac{1}{3} = 8 \times \frac{3}{4} = 6 \text{ hours to paint one sign}$$

Barry can paint two signs in 12 hours.

	Time to paint two signs in hours	Fractional part of job done in 1 hour	Fractional part each does in:		
			3 hours	$\frac{1}{2}$ hour	x hours
Bill	16	$\frac{1}{16}$	$3\left(\frac{1}{16}\right)$	$\frac{1}{2}\left(\frac{1}{16}\right)$	$x\left(\frac{1}{16}\right)$
Bob	8	$\frac{1}{8}$	$3\left(\frac{1}{8}\right)$	$\frac{1}{2}\left(\frac{1}{8}\right)$	—
Barry	12	$\frac{1}{12}$	$3\left(\frac{1}{12}\right)$	—	—

Equation: The total of all fractional parts of the job done equals 1.

$$\frac{3}{16} + \frac{3}{8} + \frac{3}{12} + \frac{1}{32} + \frac{1}{16} + \frac{x}{16} = 1$$

Multiply by the LCD, 96.

$$18 + 36 + 24 + 3 + 6 + 6x = 96$$

$$6x = 9$$

$$x = 1\frac{1}{2} \quad \text{Answer}$$

10. Let x = number of hours Joe and Russ took to finish the job

	Total time in hours	Fractional part of job done in 1 hour	Fractional part each person does in:			
			2 hours	2 hours	$\frac{1}{2}$ hour	x hours
Keith	8	$\frac{1}{8}$	$2\left(\frac{1}{8}\right)$	$2\left(\frac{1}{8}\right)$	—	—
Joe	10	$\frac{1}{10}$	—	$2\left(\frac{1}{10}\right)$	$2\left(\frac{1}{10}\right)$	$x\left(\frac{1}{10}\right)$
Russ	12	$\frac{1}{12}$	—	—	—	$x\left(\frac{1}{12}\right)$

Equation: The total of all fractional parts of the job equals 1.

$$2\left(\frac{1}{8}\right) + 2\left(\frac{1}{8}\right) + 2\left(\frac{1}{10}\right) + \frac{1}{2}\left(\frac{1}{10}\right) + x\left(\frac{1}{10}\right) + x\left(\frac{1}{12}\right) = 1$$

$$\frac{1}{4} + \frac{1}{4} + \frac{1}{5} + \frac{1}{20} + \frac{x}{10} + \frac{x}{12} = 1$$

Multiply by the LCD, 60, to clear fractions.

$$15 + 15 + 12 + 3 + 6x + 5x = 60$$

$$11x = 15$$

$$x = 1\frac{4}{11} \quad \text{Answer}$$

11. Let x = number of vines Jay and Debbie can prune together in 1 hour

	Total time to do job in hours	Fractional part of job done in 1 hour
Jay	50	$\dfrac{1}{50}$
Debbie	40	$\dfrac{1}{40}$
Together	x	$\dfrac{1}{x}$

Equation: The fractional parts of the job each can do in 1 hour equals the fractional part they can do together.

$$\frac{1}{50} + \frac{1}{40} = \frac{1}{x}$$

Multiply by the LCD, 200x, to clear fractions.

$$4x + 5x = 200$$

$$9x = 200$$

$$x = 22\frac{2}{9} \quad \textit{Answer}$$

12. Let x = number of minutes Cheryl took to finish first car

	Total time in minutes	Fractional part of job done in 1 minute	Fractional part each does in: 2 minutes	1 minute	x minutes
Rudy	10	$\dfrac{1}{10}$	$2\left(\dfrac{1}{10}\right)$	$\dfrac{1}{10}$	—
Cheryl	12	$\dfrac{1}{12}$	$2\left(\dfrac{1}{12}\right)$	$\dfrac{1}{12}$	$x\left(\dfrac{1}{12}\right)$
Tom	8	$\dfrac{1}{8}$	$2\left(\dfrac{1}{8}\right)$	—	—
Pat	15	$\dfrac{1}{15}$	$2\left(\dfrac{1}{15}\right)$	—	—

The sum of the fractional parts of the job done equals 1.

$$2\left(\frac{1}{10}\right) + 2\left(\frac{1}{12}\right) + 2\left(\frac{1}{8}\right) + 2\left(\frac{1}{15}\right) + \frac{1}{10} + \frac{1}{12} + \frac{x}{12} = 1$$

$$\frac{1}{5} + \frac{1}{6} + \frac{1}{4} + \frac{2}{15} + \frac{1}{10} + \frac{1}{12} + \frac{x}{12} = 1$$

Multiply by the LCD, 60, to clear fractions.

$$12 + 10 + 15 + 8 + 6 + 5 + 5x = 60$$

$$5x = 4$$

$$x = \frac{4}{5} \qquad \textit{Answer}$$

Chapter 9

Plane Geometric Figures

Here are a few facts to remember about geometric figures.

1. The perimeter of a rectangle equals two widths plus two lengths.
2. The sum of the three angles of any triangle equals 180°.
3. A square has four equal sides.
4. The area of a rectangle equals its length times its width.

EXAMPLE I

The length of a rectangle is equal to twice the width. The perimeter is 138 feet. What are the dimensions of the rectangle?

Solution

Let x = width of rectangle in feet (smaller number)
$\quad 2x$ = length of rectangle in feet

Here we can again make a sketch,

141

Equation: The perimeter equals two lengths plus two widths.

$$2x + 2x + x + x = 138 \quad \text{or} \quad 2(2x) + 2(x) = 138$$

$$6x = 138$$

$$\left. \begin{array}{l} x = 23 \\ 2x = 46 \end{array} \right\} \text{Answers}$$

EXAMPLE 2

A rectangle has a length which is 4 feet less than three times the width. The perimeter is 224 feet. What are the dimensions of the rectangle?

Solution

Let $\quad x$ = width of rectangle in feet
$3x - 4$ = length of rectangle in feet (4 feet less than three times the width)

Sketch:

Equation: The perimeter equals two lengths plus two widths.

$$2(x) + 2(3x - 4) = 224$$

$$2x + 6x - 8 = 224$$

$$8x = 232$$

$$\left. \begin{array}{l} x = 29 \\ 3x - 4 = 83 \end{array} \right\} \text{Answers}$$

EXAMPLE 3

The first angle of a triangle is twice the second, and the third is 5° larger than the first. Find the three angles.

Solution

Let x = number of degrees in second angle (smallest)
    $2x$ = number of degrees in first angle
    $2x + 5$ = number of degrees in third angle

Sketch:

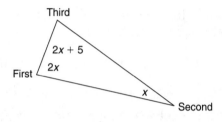

Equation: The sum of three angles of a triangle equals 180°.

$$x + 2x + (2x + 5) = 180$$

$$5x + 5 = 180$$

$$5x = 175$$

$$x = 35$$
$$2x = 70 \quad \Big\} \quad Answers$$
$$2x + 5 = 75$$

EXAMPLE 4

The length of one rectangle is two times the width. If the length is decreased by 5 feet and the width is increased by 5 feet, the area is increased by 75 square feet. Find the dimensions of the original rectangle.

143

Solution

Here we have two rectangles. Let's draw them.

2x		2x − 5	
x	Area (2x) (x) = 2x²	x + 5	Area (2x − 5) (x + 5)

Let x = width of first rectangle in feet
$2x$ = length of first rectangle in feet

Equation: The area of the first rectangle plus 75 equals the area of the second rectangle.

$$(2x)(x) + 75 = (2x - 5)(x + 5)$$

$$2x^2 + 75 = 2x^2 + 5x - 25$$

The $2x^2$ terms cancel when all x's are transposed to the left.

$$-5x = -100$$

$$\left. \begin{array}{l} x = 20 \\ 2x = 40 \end{array} \right\} \quad Answers$$

EXAMPLE 5

The first side of a triangle is 2 inches less than twice the second side. The third side is 10 inches longer than the second side. If the perimeter is 12 feet, find the length of each side.

Solution

Notice that the first and third sides are longer than the second. So we let x equal the second side. Also, note that the perimeter is given in *feet* while the sides are given in *inches*. The perimeter must be changed to the same units (inches), and 12 feet equals 144 inches.

Let x = second side in inches
$2x - 2$ = first side in inches
$x + 10$ = third side in inches

Sketch:

x $x + 10$

$2x - 2$

Equation:

$$x + (2x - 2) + (x + 10) = 144$$
$$4x + 8 = 144$$
$$4x = 136$$
$$\left.\begin{array}{l} x = 34 \\ 2x - 2 = 66 \\ x + 10 = 44 \end{array}\right\} \quad \textit{Answers}$$

Supplementary Plane Geometric Figure Problems

1. The length of a rectangular soccer field is 10 feet more than twice the width. The perimeter is 320 feet. Find the length and width.

2. The second angle of a triangle is 20° greater than the first angle. The third angle is twice the second. Find the three angles.

3. A farmer wishes to fence a rectangular area behind his barn. The barn forms one end of the rectangle, and the length of the rectangle is three times the width. How many linear feet of fence must the farmer buy if the perimeter of the rectangle is 320 feet?

4. The first side of a triangle is twice the second, and the third side is 20 feet less than three times the second. The perimeter of the triangle is 106 feet. Find the three sides.

5. The length of a room is 8 feet more than twice the width. If it takes 124 feet of molding to go around the perimeter of the room, what are the room's dimensions?

6. The perimeter of a triangular lawn is 42 yards. The first side is 5 yards less than the second, and third side is 2 yards less than the first. What is the length of each side of the lawn?

7. One side of a rectangular metal plate is five times as long as the other side. If the perimeter is 72 meters, what is the length of the shorter side of the plate?

8. A rectangular box of cereal contains 252 cubic inches of product. It is 12 inches tall and 7 inches wide. How deep is the box? (Volume equals height times width times depth.)

9. The length of a rectangle is 8 feet more than the width. If the width is increased by 4 feet and the length is decreased by 5 feet, the area of the two rectangles remains the same. Find the dimensions of the original rectangle.

Solutions to Supplementary Plane Geometric Figure Problems

1. Let $\quad x$ = width of field in feet
$2x + 5$ = length of field in feet

$2x + 10$

$x \qquad\qquad x$

$2x + 10$

Equation: The perimeter equals two lengths plus two widths.

$$320 = 2(x) + 2(2x + 10)$$
$$= 2x + 4x + 20$$
$$= 6x + 20$$
$$300 = 6x$$
$$x = 50 \quad\Big\}$$
$$2x + 10 = 110 \quad\Big\}\;\textit{Answers}$$

Check:

$$2(50) + 2(110) = 320$$
$$100 + 220 = 320$$

2. The first angle is the smallest.

Let $\qquad x$ = number of degrees in first angle
$x + 20$ = number of degrees in second angle
$2(x + 20)$ = number of degrees in third angle

146

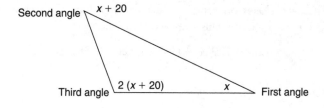

Second angle $x + 20$

Third angle $2(x + 20)$

x First angle

Equation:

$$x + (x + 20) + 2(x + 20) = 180$$
$$4x + 60 = 180$$
$$4x = 120$$
$$x = 30$$
$$x + 20 = 50 \quad \Big\} \quad \text{Answers}$$
$$2(x + 20) = 100$$

Check:

$$30 + 50 + 100 = 180$$
$$180 = 180$$

3. Let x = width of fenced area in feet
$3x$ = length of fenced area in feet

$3x$

$3x$

x

Equation:

$$2(3x) + 2(x) = 320$$
$$8x = 320$$
$$x = 40$$
$$3x = 120$$

The barn closes one end of the area. Therefore, the farmer needs to buy two lengths plus one width of fencing or

$$120 + 120 + 40 = 280 \text{ feet} \qquad Answer$$

Check:

$$40 + 40 + 120 + 120 = 320$$
$$320 = 320$$

4. The second side is the smallest.

Let $\qquad x$ = second side in feet
$2x$ = first side in feet
$3x - 20$ = third side in feet

Equation:

$$x + 2x + (3x - 20) = 106$$
$$6x - 20 = 106$$
$$6x = 126$$
$$x = 21$$
$$2x = 42 \qquad \left. \right\} \quad Answers$$
$$3x - 20 = 43$$

Check:

$$21 + 42 + 43 = 106$$
$$106 = 106$$

5. The width is smaller.

Let x = width in feet
    $2x + 8$ = length in feet (Eight feet more than twice the width equals the length.)

Equation: The perimeter equals two lengths plus two widths.

$$2(x) + 2(2x + 8) = 124$$

$$6x + 4x + 16 = 124$$

$$6x + 16 = 124$$

$$6x = 108$$

$$x = 18$$
$$2x + 8 = 44$$

} *Answers*

Check:

$$2(18) + 2(44) = 124$$

$$36 + 88 = 124$$

$$124 = 124$$

6. The second side is the largest.

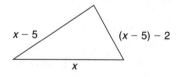

Let x = second side of lawn in yards
      $x - 5$ = first side of lawn in yards
  $(x - 5) - 2$ = third side of lawn in yards

Equation: The sum of the three sides equals the perimeter.

$$x + (x - 5) + (x - 5 - 2) = 42$$

$$x + x - 5 + x - 7 = 42$$

$$3x - 12 = 42$$

$$3x = 54$$

$$x = 18$$

$$x - 5 = 13 \quad \rbrace \ \textit{Answers}$$

$$x - 5 - 2 = 11$$

Check:

$$11 + 13 + 18 = 42$$

$$42 = 42$$

7. Let x = width of plate in feet
    $5x$ = length of plate in feet

5x

x   x

5x

Equation:

$$2(5x) + 2(x) = 72$$

$$12x = 72$$

$$x = 6$$

$$5x = 30 \quad \rbrace \ \textit{Answers}$$

Check:

$$2(6) + 2(30) = 72$$
$$72 = 72$$

8. Let x = depth of box in inches

12

7

x

Equation:

$$7(12)x = 252$$
$$84x = 252$$
$$x = 3 \quad \textit{Answer}$$

Check:

$$(7)(12)(3) = 252$$
$$252 = 252$$

9. Let x = width of original rectangle in feet
$x + 8$ = length of original rectangle in feet

$x + 8$

Original rectangle	Area $x(x + 8)$	x

Then

$$x + 4 = \text{width of new rectangle}$$

$$(x + 8) - 5 = \text{length of new rectangle}$$

Equation: The area of first rectangle, $x(x + 8)$, equals the area of the second rectangle, $(x + 4)(x + 3)$.

$$x(x + 8) = (x + 4)(x + 3)$$

$$x^2 + 8x = x^2 + 7x + 12$$

$$8x - 7x = 12$$

$$x = 12$$
$$x + 8 = 20$$ } *Answers*

Check:

$$12(20) = (12 + 4)(12 + 3)$$

$$240 = 240$$

Chapter 10

Digits

Digit problems are problems involving numbers with two or more digits. In order to understand them, you must understand what each digit in a number means. For example, the number 329 means

3 one hundreds	or	$3 \times 100 = 300$
2 tens	or	$2 \times 10 = 20$
9 units (ones)	or	$9 \times 1 = \underline{9}$
	Sum	$= 329$

If you have a digit problem in algebra, you have to express the number as above. For example, if the tens digit is twice the units (ones) digit, we can express the number:

Let x = units digit
$2x$ = tens digit

Then if we multiply the tens digit by 10 and add it to the units digit, we get

$$\text{Number} = 10(2x) + x$$

Often a digit problem involves a *reversed* number; that is, the digits are interchanged. The numbers 39 and 93 have digits reversed. We can express these two numbers as follows: $(10 \times 3) + 9$ and $(10 \times 9) + 3$.

Now take the number above, $10(2x) + x$. To reverse it, we reverse the units digit x and the tens digit $2x$. We get $10(x) + 2x$.

EXAMPLE I

The tens digit of a certain number is 3 less than the units digit. The sum of the digits is 11. What is the number?

Solution

Let x = units digit
    $x - 3$ = tens digit

Here there is no need to write the number itself except for the answer.

Equation:

$$x + (x - 3) = 11$$

$$2x = 14$$

$$x = 7 \quad \text{(units digit)}$$

$$x - 3 = 4 \quad \text{(tens digit)}$$

Using the values above for units and tens, we find the number if $(4 \times 10) + 7$ or 47. *Answer*

EXAMPLE 2

The tens digit of a number is twice the units digit. If the digits are reversed, the new number is 27 less than the original. Find the original number.

Solution

Let x = units digit
  $2x$ = tens digit

Then the original number is $10(2x) + x$, the reversed number is $10(x) + 2x$, and the new number is the original number less 27.

Equation:

$$10(x) + 2x = 10(2x) + x - 27$$

$$12x = 21x - 27$$

$$-9x = -27$$

$$x = 3 \quad \text{(units digit)}$$

$$2x = 6 \quad \text{(tens digit)}$$

The number is $(6 \times 10) + 3$ or 63. *Answer*

EXAMPLE 3

The sum of the digits in a two-digit number is 12. If the digits are reversed, the number is 18 greater than the original number. What is the number?

Solution

Let $\quad x = $ units digit
$12 - x = $ tens digit

Then the original number is $10(12 - x) + x$ and the reversed number is $10(x) + (12 - x)$.

Equation: The reversed number is the original number plus 18.

$$10(x) + (12 - x) = 10(12 - x) + (x) + 18$$

$$9x + 12 = 120 - 10x + x + 18$$

$$= -9x + 138$$

$$18x = 126$$

$$x = 7 \quad \text{(units digit)}$$

$$12 - x = 5 \quad \text{(tens digit)}$$

The number is $(5 \times 10) + 7$ or 57. *Answer*

Check:

$$10(7) + (12 - 7) = 10(12 - 7) + (7) + 18$$

$$70 + 5 = 50 + 7 + 18$$

$$75 = 75$$

Supplementary Digit Problems

1. The tens digit of a certain number is 5 more than the units digit. The sum of the digits is 9. Find the number.
2. The tens digit of a two-digit number is twice the units digit. If the digits are reversed, the new number is 36 less than the original number. Find the number.
3. The sum of the digits of a two-digit number is 13. The units digit is 1 more than twice the tens digit. Find the number.
4. The sum of the digits of a three-digit number is 6. The hundreds digit is twice the units digit, and the tens digit equals the sum of the other two. Find the number.
5. The units digit is twice the tens digit. If the number is doubled, it will be 12 more than the reversed number. Find the number.
6. Eight times the sum of the digits of a certain two-digit number exceeds the number by 19. The tens digit is 3 more than the units digit. Find the number.
7. The ratio of the units digit to the tens digit of a two-digit number is one-half. The tens digit is 2 more than the units digit. Find the number.
8. There is a two-digit number whose units digit is 6 less than the tens digit. Four times the tens digit plus five times the units digit equals 51. Find the digits.
9. The tens digit is 2 less than the units digit. If the digits are reversed, the sum of the reversed number and the original number is 154. Find the original number.
10. A three-digit number has a tens digit 2 greater than the units digit and a hundreds digit 1 greater than the tens digit. The sum of the tens and hundreds digits is three times the units digit. What is the number?

Solutions to Supplementary Digit Problems

1. Let $\quad x =$ units digit
   $x + 5 =$ tens digit

   Equation:

   $$x + (x + 5) = 9$$
   $$2x + 5 = 9$$

$$2x = 4$$

$$x = 2 \quad \text{(units digit)}$$

$$x + 5 = 7 \quad \text{(tens digit)}$$

The number is $(7 \times 10) + 2$ or 72.   *Answer*

2. Let x = units digit
   $2x$ = tens digit

Then the number is $10(2x) + x$ and the reversed number is $10(x) + 2x$.

Equation:

$$10(x) + 2x = 10(2x) + x - 36$$

$$12x = 21x - 36$$

$$-9x = -36$$

$$x = 4 \quad \text{(units digit)}$$

$$2x = 8 \quad \text{(tens digit)}$$

The number is $(8 \times 10) + 4$ or 84. *Answer*

3. Let x = units digit
   $13 - x$ = tens digit

Equation: The units digit is twice the tens digit plus 1.

$$x = 2(13 - x) + 1$$

$$= 26 - 2x + 1$$

$$3x = 27$$

$$x = 9 \quad \text{(units digit)}$$

$$13 - x = 4 \quad \text{(tens digit)}$$

The number is $(4 \times 10) + 9$ or 49.   *Answer*

4. Let x = units digit
   $2x$ = hundreds digit
   $x + 2x$ = tens digit

Equation:

$$x + 2x + (x + 2x) = 6$$

$$6x = 6$$

$$x = 1 \text{ (units digit)}$$

$$2x = 2 \text{ (hundreds digit)}$$

$$x + 2x = 3 \text{ (tens digit)}$$

The number is $(2 \times 100) + (3 \times 10) + 1$ or 231. *Answer*

5. Let x = tens digit
     $2x$ = units digit

Then the number is $10(x) + 2x$ and the reversed number is $10(2x) + x$.

Equation: Two times the number equals 12 more than the reversed number.

$$2[10(x) + (2x)] = 10(2x) + x + 12$$

$$2(12x) = 21x + 12$$

$$24x = 21x + 12$$

$$3x = 12$$

$$x = 4 \quad \text{(tens digit)}$$

$$2x = 8 \quad \text{(units digit)}$$

The number is $(4 \times 10) + 8$ or 48. *Answer*

6. Let x = units digit (smaller)
       $x + 3$ = tens digit

Then the number is $10(x + 3) + x$.

Equation: Eight times the sum of the digits exceeds the number by 19.

$$8[x + (x + 3)] - [10(x + 3) + x] = 19$$

$$16x + 24 - 10x - 30 - x = 19$$

$$5x - 6 = 19$$

$$5x = 25$$

$$x = 5 \quad \text{(units digit)}$$

$$x + 3 = 8 \quad \text{(tens digit)}$$

The number is $(8 \times 10) + 5$ or 85.   *Answer*

7. Let $x =$ units digit
       $x + 2 =$ tens digit

Equation: The ratio is a fractional relationship.

$$\frac{x}{x + 2} = \frac{1}{2}$$

Multiply by the LCD, $2(x + 2)$.

$$2x = x + 2$$

$$x = 2 \quad \text{(units digit)}$$

$$x + 2 = 4 \quad \text{(tens digit)}$$

The number is $(4 \times 10) + 2$ or 42.   *Answer*

8. Let $x =$ units digit
       $x + 6 =$ tens digit

Equation: Four times the tens digit plus five times the units digit equal 51.

$$4(x + 6) + 5x = 51$$

$$4x + 24 + 5x = 51$$

$$9x + 24 = 51$$

$$9x = 27$$

$$x = 3 \quad \text{(units digit)}$$

$$x + 6 = 9 \quad \text{(tens digit)}$$

The number is $(9 \times 10) + 3$ or 93.   *Answer*

9. Let $x =$ units digit
       $x - 2 =$ tens digit

Then the number is $10(x - 2) + x$ and the reversed number is $10(x) + (x - 2)$.

159

Equation: The reversed number plus the original number equal 154.

$$10(x) + (x - 2) + 10(x - 2) + x = 154$$
$$10x + x - 2 + 10x - 20 + x = 154$$
$$22x - 22 = 154$$
$$22x = 176$$
$$x = 8 \quad \text{(units digit)}$$
$$x - 2 = 6 \quad \text{(tens digit)}$$

The number is $(6 \times 10) + 8$ or 68. *Answer*

10. Let
$$x = \text{units digit}$$
$$x + 2 = \text{tens digit}$$
$$(x + 2) + 1 = \text{hundreds digit}$$

Equation: The sum of the tens and hundreds digits is three times the units digit.

$$(x + 2) + (x + 2) + 1 = 3x$$
$$2x + 5 = 3x$$
$$-x = -5$$
$$x = 5 \quad \text{(units digit)}$$
$$x + 2 = 7 \quad \text{(tens digit)}$$
$$(x + 2) + 1 = 8 \quad \text{(hundreds digit)}$$

The number is $(8 \times 100) + (7 \times 10) + 5$ or 875. *Answer*

Chapter 11

Solutions Using Two Unknowns

This chapter will repeat examples of problems which were worked in earlier chapters. This time they will be worked using two unknowns (x and y) and two equations solved simultaneously. A few word problems are more easy to work with one unknown, but many become easier when two unknowns are used. The same procedures that we learned in the problems with one unknown apply here also. You may find it easier to set up a problem with two unknowns because representing the second unknown is so easy. The first will be represented by x and the second by y. Remember, *values must be paired at the end of the problem*. The value of x alone is not enough.

The problems which follow are grouped to correspond to the appropriate earlier chapters. Examples of most types of problems will be worked again with two unknowns. Occasionally, a certain type of problem is more difficult to work with two unknowns.

Number Problems

EXAMPLE 1

The sum of two numbers is 41. The larger number is 1 less than twice the smaller. Find the numbers.

Solution

The last sentence tells us we have two unknowns. Now we shall use x and y to represent these unknowns. The two statements in the problem give us information for two equations.

Let x = smaller number
y = larger number

Equations: The sum of two numbers is 41.

$$x + y = 41 \tag{1}$$

The larger is 1 less than twice the smaller.

$$y = 2x - 1 \tag{2}$$

Solve the two equations together. Substitute the value of y in Eq. (2) for y in Eq. (1) because Eq. (2) is already solved for y, making substitution easy. If one equation or the other is not solved for x or y, solve the simplest one and then substitute in the other equation.

$$x + (2x - 1) = 41 \tag{1}$$

$$3x - 1 = 41$$

$$3x = 42$$

$$x = 14$$

Substituting the value of x in Eq. (2),

$$y = 2x - 1 \tag{2}$$

$$= 2(14) - 1$$

$$\left. \begin{array}{l} y = 27 \\ x = 14 \end{array} \right\} \text{\emph{Answers}}$$

When you use the substitution method, remember to substitute the value of the first unknown in the equation which was used in the substitution step to get the second unknown. In this problem you substitute the value of x back into $y = 2x - 1$.

EXAMPLE 2

There are two numbers whose sum is 53. Three times the smaller number is equal to 19 more than the larger number. What are the numbers?

Solution

Let x = smaller number
y = larger number

Equations: Two numbers whose sum is 53.

$$x + y = 53 \tag{1}$$

Three times the smaller equals 19 more than the larger.

$$3x = y + 19 \tag{2}$$

After subtracting y from both sides of Eq. (2), we get

$$3x - y = 19$$

Adding Eq. (1) to the last equation to eliminate y, we get

$$4x = 72$$

$$x = 18$$

Substituting the value of x in Eq. (1),

$$x + y = 53 \tag{1}$$

$$18 + y = 53$$

$$\left. \begin{array}{c} y = 35 \\ x = 18 \end{array} \right\} \quad \textit{Answers}$$

Time, Rate, and Distance Problems

EXAMPLE I

Kevin flies (in his private plane) from Los Angeles (LA) to San Francisco (SF), a distance of 450 miles, in 3 hours. He flies back in $2\frac{7}{19}$ hours. If the wind is blowing from the north at a steady rate during both flights, find the airspeed of the plane and the velocity of the wind.

Read the problem carefully. There are two unknowns. There are two statements of relationship, one for each equation. Determine what you are trying to find. You always represent the two unknowns by two different letters in this type of solution. The basic procedures are the same as with one unknown. Draw the sketch.

Solution

Let x = airspeed of plane in mph (in still air)
y = velocity of wind

	Time	Rate	Distance
LA to SF	3	$x - y$	$3(x - y)$
SF to LA	$2\frac{7}{19}$	$x + y$	$2\frac{7}{19}(x + y)$

Here there are *three* possible equations. You can use any two.

$$3(x - y) = 450 \tag{1}$$

$$2\frac{7}{19}(x + y) = 450 \tag{2}$$

$$3(x - y) = 2\frac{7}{19}(x + y) \tag{3}$$

It is more convenient to use Eqs. (1) and (2) because the unknowns are on the same side of the equations (proper format for equations in two unknowns). Simplifying Eq. (2),

$$2\frac{7}{19}(x + y) = 450 = \frac{45}{19}(x + y)$$

Clear parentheses.

$$3x - 3y = 450 \tag{1}$$

$$\frac{45x + 45y}{19} = 450 \tag{2}$$

Clear the fraction in Eq. (2).

$$45x + 45y = 8550 \tag{2}$$

Divide Eq. (1) by 3 and Eq. (2) by 45.

$$x - y = 150 \tag{1}$$

$$x + y = 190 \tag{2}$$

Add the equations (to cancel the y's) and solve for x.

$$2x = 340$$

$$x = 170 \text{ mph} \quad \text{(airspeed)} \quad \textit{Answer}$$

Substitute the value of x in Eq. (2).

$$x + y = 190 \tag{2}$$

$$170 + y = 190$$

$$y = 20 \text{ mph} \quad \text{(wind velocity)} \quad \textit{Answer}$$

Mixture Problems

EXAMPLE I

A farmer has 100 gallons of 70% pure disinfectant. He wishes to mix it with disinfectant which is 90% pure to obtain 75% pure disinfectant. How much of the 90% disinfectant must he use?

Solution

There are two unknowns. Either one may be labeled x and the other y. Note that we only have to find one in this problem.

Let x = number of gallons of 90% disinfectant
  y = number of gallons of 75% disinfectant

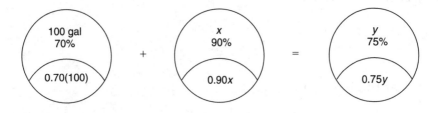

Equations:

$$100 + x = y \quad \text{(total gallons)} \tag{1}$$

$$0.70(100) + 0.90x = 0.75y \quad \text{(total disinfectant)} \tag{2}$$

Simplifying Eq. (2),

$$70 + 0.90x = 0.75y$$

$$7000 + 90x = 75y$$

Then divide by 5.

$$1400 + 18x = 15y \tag{2}$$

Substitute Eq. (1) in Eq. (2).

$$1400 + 18x = 15(100 + x)$$

$$= 1500 + 15x$$

$$3x = 100$$

$$x = 33\frac{1}{3} \text{ gallons} \qquad \textit{Answer}$$

EXAMPLE 2

A farmer mixes milk containing 3% butterfat with cream containing 30% butterfat to obtain 900 gallons of milk which is 8% butterfat. How much of each must the farmer use?

Solution

Let x = number of gallons of milk
  y = number of gallons of cream

166

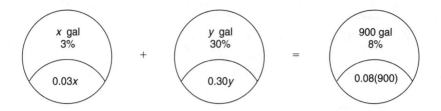

Equations:

$$x + y = 900 \tag{1}$$

$$0.03x + 0.30y = 0.08(900) \tag{2}$$

Simplifying Eq. (2),

$$0.03x + 0.30y = 72$$

Multiply by 100 to clear decimals.

$$3x + 30y = 7200$$

Then divide Eq. (2) by 3.

$$x + 10y = 2400 \tag{2}$$

Subtract Eq. (1) from Eq. (2) to eliminate x.

$$9y = 1500$$

$$y = 166\frac{2}{3} \text{ gallons of 30\% butterfat} \quad \textit{Answer}$$

Substitute y in Eq. (1).

$$x = 733\frac{1}{3} \text{ gallons of 3\% butterfat} \quad \textit{Answer}$$

Coin Problems

EXAMPLE I

Ezekiel has some coins in his pocket consisting of dimes, nickels, and pennies. He has two more nickels than dimes and three times as many pennies as nickels. How many of each kind of coin does he have if the total value is 52 cents?

167

Solution

Let x = number of dimes
      y = number of nickels
      z = number of pennies

Equations:

$$y = x + 2 \qquad (1)$$

$$z = 3y \qquad (2)$$

$$10x + 5y + x = 52 \qquad (3)$$

Remember to convert x, y, and z to cents in Eq. (3). Then substitute Eqs. (1) and (2) in Eq. (3) so all are in terms of y.

$$10(y - 2) + 5y + (3y) = 52 \qquad (3)$$

$$18y = 72$$

$$y = 4 \text{ nickels}$$

$$z = \quad 3y = 3(4) \quad = 12 \text{ pennies}$$

$$x = y - 2 = 4 - 2 = 2 \text{ dimes}$$

Answers

EXAMPLE 2

Mrs. Delahoyd received a total of 180 coins each day for her cash drawer at the restaurant where she was a cashier. She had the same number of nickels as quarters and the same number of pennies as dimes. She had 10 more dimes than nickels. How many of each did she have?

Solution

Let P = number of pennies
      N = number of nickels
      D = number of dimes
      Q = number of quarters

Equations:

$$P + N + D + Q = 180 \qquad (1)$$

$$N = Q \tag{2}$$

$$P = D \tag{3}$$

$$D = N + 10 \tag{4}$$

We need to find two equations which will each have the same two unknowns. First, substitute Eqs. (2) and (3) in Eq. (1) to make an equation with values in Q and D.

$$(D) + (Q) + D + Q = 180 \tag{1}$$

$$2D + 2Q = 180$$

Divide by 2.

$$D + Q = 90 \tag{1}$$

Second, substitute Eq. (2) into Eq. (4) to get a second equation with values in Q and D.

$$D = (Q) + 10 \tag{4}$$

Solve Eqs. (1) and (4) together.

$$D + Q = 90 \tag{1}$$

$$D - Q = 10 \tag{4}$$

Add Eqs. (1) and (4) to eliminate Q.

$$2D = 100$$

$$D = 50$$

Substitute this value for D in Eq. (1) (where D and Q are the only unknowns) to find Q.

$$D + Q = 90 \tag{1}$$

$$50 + Q = 90$$

$$Q = 40$$

Substitute this value for Q in Eq. (2) to find N.

$$N = Q \tag{2}$$

$$N = 40$$

Substitute the value for D in Eq. (3) to find P.

$$P = D \tag{3}$$

$$\left.\begin{array}{l} P = 50 \\ D = 50 \\ Q = 40 \\ N = 40 \end{array}\right\} \quad \textit{Answers}$$

Age Problems

EXAMPLE I

A man is four times as old as his son. In 3 years the father will be three times as old as his son. How old is each now?

Solution

Let x = son's age now
        y = father's age now
  $x + 3$ = son's age in 3 years
  $y + 3$ = father's age in 3 years

Equations: The father's age is four times his son's age *now*.

$$y = 4x \tag{1}$$

The father's age is three times his son's age *then*.

$$y + 3 = 3(x + 3) \tag{2}$$

Substitute Eq. (1) in Eq. (2).

$$(4x) + 3 = 3(x + 3) \tag{2}$$

$$4x + 3 = 3x + 9$$

$$x = 6$$

Substituting the value for x in Eq. (1),

$$\left.\begin{array}{r} y = 24 \\ x = 6 \end{array}\right\} \text{ \textit{Answers}}$$

EXAMPLE 2

Bettina's age is three times Melvina's. If 20 is added to Melvina's age and 20 is subtracted from Bettina's, their ages will then be equal. How old is each now?

Solution

Let x = Bettina's age now
  y = Melvina's age now

Equations: Bettina's age is three times Melvina's.

$$x = 3y \tag{1}$$

If 20 is added to Melvina's age and 20 is subtracted from Bettina's, their ages are equal.

$$y + 20 = x - 20 \tag{2}$$

Substitute Eq. (1) in Eq. (2).

$$y + 20 = (3y) - 20 \tag{2}$$
$$-2y = -40$$
$$y = 20$$

Substitute y in Eq. (1).

$$x = 3y \tag{1}$$
$$= 3(20)$$
$$\left.\begin{array}{r} x = 60 \\ y = 20 \end{array}\right\} \text{ \textit{Answers}}$$

Lever Problems

Lever problems are easier in one unknown. In general, you will find some problems easier to work in one unknown and some easier in two unknowns. There is no rule to follow.

Finance Problems

EXAMPLE I

Mr. McGregor invested $10,000, part at 5 percent and part at 9 percent. If the total yearly interest was $660, how much did he invest at each rate?

Solution

Let x = amount in dollars invested at 5 percent
y = amount in dollars invested at 9 percent
Then

$$0.05x = \text{interest on 5 percent investment}$$

$$0.09y = \text{interest on 9 percent investment}$$

Equations:

$$0.05x + 0.09y = 660 \qquad \text{(total \textit{interest} equals \$660)} \qquad (1)$$

$$x + y = 10{,}000 \qquad \text{(total \textit{investment} equals \$10,000)} \qquad (2)$$

Multiply Eq. (1) by 100 and Eq. (2) by 5.

$$5x + 9y = 66{,}000 \qquad (1)$$

$$5x + 5y = 50{,}000 \qquad (2)$$

Subtract to eliminate x.

$$4y = 16{,}000$$

$$y = 4000$$

Substitute the value for y in Eq. (2).

$$x + 4000 = 10{,}000 \qquad (2)$$

$$x = 6000$$

Always pair x and y values to show the solution set.

$$\left. \begin{array}{l} x = 6000 \\ y = 4000 \end{array} \right\} \quad \textit{Answers}$$

EXAMPLE 2

Tickets for the school play sold at $2 each for adults and
75 cents each for students. If there were four times as many
adult tickets sold as students tickets, and the total receipts
were $1750, how many adult and how many student tickets
were sold?

Solution

Let x = number of adult tickets sold at $2 each
      y = number of student tickets sold at 75 cents each
Then

$2x$ = total number of dollars received from adult tickets

$0.75y$ = total number of dollars received from student tickets

Equations: Four times as many adult tickets as student
tickets were sold.

$$x = 4y \tag{1}$$

Total receipts were $1750.

$$2x + 0.75y = 1750 \tag{2}$$

Clear decimals in Eq. (2).

$$200x + 75y = 175{,}000 \tag{2}$$

Divide Eq. (2) by constant 25 to simplify the equation.

$$8x + 3y = 7000 \tag{2}$$

Substitute the value of x in Eq. (1) into Eq. (2) so that Eq. (2)
will be in terms of y.

$$8(4y) + 3y = 7000 \tag{2}$$

$$35y = 7000$$

$$y = 200$$

Substitute the value of y in Eq. (1).

$$x = 4y \qquad\qquad (1)$$

$$= 4(200)$$

$$\left.\begin{array}{l} x = 800 \\ y = 200 \end{array}\right\} \text{ Answers}$$

Work Problems

Work problems are more easily solved with one unknown.

Plane Geometric Figure Problems

EXAMPLE I

The length of a rectangle is 5 feet more than twice the width. The perimeter is 28 feet. Find the length and width of the rectangle.

Solution

Let x = width in feet
  y = length in feet

Equations: The length is 5 feet more than twice the width.

$$y = 2x + 5 \qquad\qquad (1)$$

The perimeter is 28 feet.

$$2x + 2y = 28 \qquad\qquad (2)$$

Divide Eq. (2) by constant 2 to simplify.

$$x + y = 14 \qquad (2)$$

Substitute the value of y in Eq. (1) into Eq. (2) to express Eq. (2) in terms of x.

$$x + (2x + 5) = 14 \qquad (2)$$

$$3x + 5 = 14$$

$$3x = 9$$

$$x = 3$$

Substitute x in Eq. (1).

$$y = 2(3) + 5 \qquad (1)$$

$$= 6 + 5$$

$$\left. \begin{array}{l} y = 11 \\ x = 3 \end{array} \right\} \quad \textit{Answers}$$

EXAMPLE 2

The second angle of a triangle is 20° greater than the first angle. The third angle is twice the second. Find the three angles

Solution

Let x = number of degrees in first angle
y = number of degrees in second angle
z = number of degrees in third angle

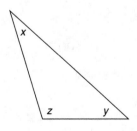

Equations: The second angle is 20° greater than the first angle.

$$y = x + 20 \tag{1}$$

The third is twice the second.

$$z = 2y \tag{2}$$

The sum of the angles of a triangle is 180°.

$$x + y + z = 180 \tag{3}$$

With three unknowns there must be three equations.

Remember, we need to get two equations in two unknowns to solve them together to eliminate one unknown. Sometimes it is possible to substitute two equations into the third so that all but one unknown is eliminated. We can solve Eq. (1) for x. Equation (2) has z in terms of y. So we can now substitute Eqs. (1) and (2) into Eq. (3) to get Eq. (3) in terms of y.

$$x = y - 20 \tag{1}$$

$$z = 2y \tag{2}$$

$$x + y + z = 180 \tag{3}$$

$$(y - 20) + y + (2y) = 180 \tag{3}$$

Simplifying,

$$4y - 20 = 180$$

$$4y = 200$$

$$y = 50$$

$$z = 2y$$

$$z = 100$$

$$x = y - 20$$

$$
\left.\begin{array}{l}
x = 30 \\
y = 50 \\
z = 100
\end{array}\right\} \quad \textit{Answers}
$$

Digit Problems

EXAMPLE I

The tens digit of a two-digit number is 5 more than the units digit. The sum of the digits is 9. Find the number.

Solution

Let x = units digit
      y = tens digit

You could let x = tens digit and y = units digit. In using two letters for the two unknowns, it does not make any difference which unknown x stands for and which unknown y stands for. You will get the correct answer either way.

The tens digit is 5 more than the units digit.

$$y = x + 5 \qquad (1)$$

The sum of the digits is 9.

$$x + y = 9 \qquad (2)$$

Because Eq. (1) is already solved for y in terms of x, it suggests substitution. You substitute $x + 5$ for y in Eq. (2).

$$x + (x + 5) = 9 \qquad (2)$$

$$2x + 5 = 9$$

$$2x = 4$$

$$x = 2$$

To find the value of y, go back to Eq. (1) which expresses y in terms of x.

$$y = x + 5 \qquad (1)$$

$$= (2) + 5$$

$$y = 7$$

$$x = 2$$

The number is 72.   *Answer*

EXAMPLE 2

The sum of the digits of a three-digit number is 6. The hundreds digit is twice the units digit. The tens digit equals the sum of the other two digits. What is the number?

Solution

Let u = units digit
  t = tens digit
  h = hundreds digit

The sum of the digits is 6.

$$u + t + h = 6 \qquad (1)$$

The hundreds digit is twice the units digit.

$$h = 2u \qquad (2)$$

The tens digit equals the sum of the other two.

$$t = u + h \qquad (3)$$

Look at the above equations to see what relationships you have. Equation (2) has h in terms of u, so see if you can find another letter in terms of u. If you substitute Eq. (2) into Eq. (3) you will have a t in terms of u.

$$t = u + (2u) \qquad (3)$$

$$= 3u$$

So you now have

$$u + t + h = 6 \qquad (1)$$

$$h = 2u \qquad (2)$$

$$t = 3u \qquad (3)$$

178

By substituting Eqs. (2) and (3) in Eq. (1), we will have Eq. (1) in terms of u.

$$u + (3u) + (2u) = 6 \tag{1}$$

$$6u = 6$$

$$u = 1$$

$$h = 2u = 2(1) = 2$$

$$t = 3u = 3(1) = 3$$

The number is 231.   *Answer*

Chapter 12

Quadratics

Certain problems of various types involve quadratic equations. However, as in solving equations with two letters representing the unknowns (Chap. 11), the basic procedures remain the same.

EXAMPLE 1

A rectangular pool is surrounded by a walk 4 feet wide. The pool is 6 feet longer than it is wide. If the total area is 272 square feet more than the area of the pool, what are the dimensions of the pool?

Solution

Let $\quad x =$ width of pool in feet
$\quad x + 6 =$ length of pool in feet

The dimensions of the outside rectangle will each be 8 feet more than the dimensions of the inside rectangle (two widths of the walk). The area of the pool is $x(x + 6)$. The area of the large rectangle (total area) is $(x + 8)(x + 14)$.

Equation: The total area is 272 square feet more than the pool area.

$$(x + 8)(x + 14) = x(x + 6) + 272$$

$$x^2 + 22x + 112 = x^2 + 6x + 272$$

$$16x = 160$$

$$\left.\begin{array}{r} x = 10 \\ x + 6 = 16 \end{array}\right\} \quad \textit{Answers}$$

EXAMPLE 2

One man can pour a cement slab in 2 hours less than it takes his competitor to do the same job. If they worked together, they could do it in $2\frac{2}{5}$ hours. How long does it take each to pour the slab alone?

Solution

This is a typical work problem, so you diagram it as in Chap. 8.

Let x = time in hours for first man
    $x + 2$ = time in hours for second man

	Total time in hours	Fractional part of job done in 1 hour
First man	x	$\dfrac{1}{x}$
Second man	$x + 2$	$\dfrac{1}{x + 2}$
Together	$2\dfrac{2}{5}$	$\dfrac{1}{2\frac{2}{5}}$

Equation: The total fractional part each can do in 1 hour equals the fractional part they can do together.

$$\frac{1}{x} + \frac{1}{x+2} = \frac{1}{2\frac{2}{5}}$$

$$\frac{1}{2\frac{2}{5}} = \frac{1}{\frac{12}{5}} = 1 \div \frac{12}{5} = \frac{5}{12}$$

$$\frac{1}{x} + \frac{1}{x+2} = \frac{5}{12}$$

Multiply by the LCD, $12x(x + 2)$, to clear fractions.

$$12(x + 2) + 12x = 5x(x + 2)$$

$$12x + 24 + 12x = 5x^2 + 10x$$

$$-5x^2 - 10x = -24x - 24$$

$$5x^2 + 10x = 24x + 24$$

$$5x^2 - 14x - 24 = 0$$

$$(5x + 6)(x + 4) = 0$$

$$x - 4 = 0$$

$$\left.\begin{array}{l} x = 4 \\ x + 2 = 6 \end{array}\right\} \quad \textit{Answers}$$

The second factor gives a negative root, which is unacceptable in a word problem.

EXAMPLE 3

Jason traveled 120 miles at an average rate of speed. If he could have increased his speed 15 mph, he could have covered the same distance in $\frac{2}{5}$ of an hour less time. How fast did he travel?

Solution

Let x = rate in mph before increase

$x + 15$ = rate in mph after increase

	Time	Rate	Distance
Before	$\dfrac{120}{x}$	x	120
After	$\dfrac{120}{x+15}$	$x+15$	120

Equation: Time *before* the increase is two-fifths of an hour more than time after.

$$\frac{120}{x} = \frac{120}{x+15} + \frac{2}{5}$$

Multiply by the LCD to clear fractions.

$$5(120)(x + 15) = 5(120)x + 2x(x + 15)$$
$$600x + 9000 = 600x + 2x^2 + 30x$$
$$2x^2 + 30x - 9000 = 0$$
$$x^2 + 15x - 4500 = 0$$
$$(x + 75)(x - 60) = 0$$
$$x + 75 = 0$$
$$x = -75 \quad \text{(drop as extraneous answer)}$$
$$x - 60 = 0$$
$$x = 60 \quad \text{mph for trip before increase}$$
$$x + 15 = 75 \quad \text{mph if speed increased}$$

Answers

Note: Remember that in quadratics you will sometimes have an extra answer which is negative and has to be discarded. When you substitute to find the second answer, it *sometimes* turns out to be a positive form of this number, but don't count on it.

EXAMPLE 4

Two consecutive even numbers have a product of 624. What are the numbers?

Solution

Let x = first consecutive even number

    $x + 2$ = second consecutive even number

Equation:

$$x(x + 2) = 624$$

$$x^2 + 2x = 624$$

$$x^2 + 2x - 624 = 0$$

$$(x + 26)(x - 24) = 0$$

$$x - 24 = 0$$

$x = 24$ (first consecutive even number)

$x + 2 = 26$ (second consecutive even number)

    Answers

EXAMPLE 5

A 60- by 80-foot rectangular walk in a park surrounds a flower bed. If the walk is of uniform width and its area is equal to the area of the flower bed, how wide is the walk?

Solution

Let x = width of walk in feet

The area of the large rectangle is $80 \times 60 = 4800$, the area of the small rectangle is one-half the area of the large rectangle = 2400, and the formula for the area of the small rectangle is $(80 - 2x)(60 - 2x)$.

Equation:

$$(80 - 2x)(60 - 2x) = 2400$$

$$4800 - 280x + 4x^2 = 2400$$

$$4x^2 - 280x + 2400 = 0$$

$$x^2 - 70x + 600 = 0$$

$$(x - 10)(x - 60) = 0$$

$$x - 10 = 0$$

$$x = 10 \quad \text{(width of walk in feet)} \qquad Answer$$

$$x - 60 = 0$$

$$x = 60 \quad \text{(width of walk in feet, but not acceptable; too large for width of sidewalk)}$$

EXAMPLE 6

A girl is 12 years older than her sister. The product of their ages is 540. How old is each?

Solution

Let $\quad x$ = sister's age

$\quad x + 12$ = girl's age

Equation:

$$x(x + 12) = 540$$

$$x^2 + 12x - 540 = 0$$

$$(x + 30)(x - 18) = 0$$

$$x - 18 = 0$$

$$\left.\begin{array}{l} x = 18 \text{ years} \quad \text{(sister's age)} \\ x + 12 = 30 \text{ years} \quad \text{(girl's age)} \end{array}\right\} \quad \textit{Answers}$$

$$x + 30 = 0$$

$$x = -30 \text{ (unacceptable)}$$

EXAMPLE 7

A number is 1 more than twice another. Their squares differ by 176. What are the numbers?

Solution

Let $\quad x = $ smaller number

$\quad 2x + 1 = $ larger number

Equation:

$$(2x + 1)^2 = x^2 + 176$$

$$4x^2 + 4x + 1 = x^2 + 176$$

$$3x^2 + 4x - 175 = 0$$

$$(3x + 25)(x - 7) = 0$$

$$x - 7 = 0$$

$$\left.\begin{array}{l} x = 7 \quad \text{(smaller number)} \\ 2x + 1 = 15 \quad \text{(larger number)} \end{array}\right\} \quad \textit{Answers}$$

EXAMPLE 8

The side of a square equals the width of a rectangle. The length of the rectangle is 6 feet longer than its width. The sum of their areas is 176 square feet. Find the side of the square.

Solution

Let $x = $ side of square

Then $x + 6$ equals the length of the rectangle, x^2 equals the area of the square, and $x(x + 6)$ equals the area of the rectangle.

x	
x	Area x^2

$x + 6$	
x	Area $x(x + 6)$

Equation:

$$x^2 + x(x + 6) = 176$$

$$x^2 + x^2 + 6x = 176$$

$$2x^2 + 6x - 176 = 0$$

$$x^2 + 3x - 88 = 0$$

$$(x + 11)(x - 8) = 0$$

$$x - 8 = 0$$

$$x = 8 \quad \text{(side of square in feet)} \quad \textit{Answer}$$

Chapter 13

Miscellaneous Problem Drill

1. One number is 11 more than three times another. Their sum is 111. What are the numbers?
2. The denominator of a fraction is 24 more than the numerator. The value of the fraction is $\frac{1}{3}$. Find the numerator and denominator.
3. There are three consecutive integers. The sum of the smallest and largest is 36. Find the integers.
4. Take a number. Double the number. Subtract 6 from the result and divide the answer by 2. The quotient will be 20. What is the number?
5. The sum of three consecutive odd numbers is 249. Find the numbers.
6. There are three consecutive even numbers such that twice the first is 20 more than the second. Find the numbers.
7. A carpenter needs to cut a 14-foot board into three pieces so that the second piece is twice as long as the first and the third is twice the second. How long is the shortest piece?
8. Twice a certain number plus three times the same number is 135. Find the number.
9. A two-digit number has a tens digit 1 greater than the units digit. The sum of the number and the number formed by reversing the digits is 77. Find the number.
10. The tens digit of a two-digit number is 5 more than the units digit. If 3 is subtracted from the number and 2 is added to the reversed number, the former will be twice the latter. What is the number?

11. Mr. Geld has $50,000 to invest. Part of it is put in the bank at 6 percent, and part he puts in a savings and loan at 9 percent. If his yearly interest (simple) is $3660, how much did he invest at each rate?

12. A store advertises a 20 percent-off sale. If an article is marked for the sale at $24.48, what is the regular price?

13. The Mountaineering Shop was owned jointly by Dave, Steven, and Pierre. Steve put up $2000 more than Dave, and Pierre owned a half interest. If the total cost of the shop was $52,000, how much did each man invest?

14. Three houses were for rent by the Jippem Realty Company. They charged $300 more per month for the second house than for the first. The third house rented for the same as the second but was vacant for 2 months for repairs. How much per month did each house rent for if the rent receipts for the year were $37,200?

15. Wolfgang and Heinrich worked as electricians for $40 and $44 per hour, respectively. One month Wolfgang worked 10 hours more than Heinrich. If their total income for the month was $11,320, how many hours did each work during the month?

16. Mike and Ike each inherited a sum of money from an uncle. Mike received $800 more than Ike. Ike invested his at 8 percent and Mike invested his at 7 percent. If Mike received $16 a year more than Ike in interest, how much did each inherit?

17. Mary needs a 50% solution of alcohol. How many liters of pure alcohol must she add to 10 liters of 40% alcohol to get the proper solution?

18. A 20% nitric acid solution and a 45% nitric acid solution are to be mixed to make 6 quarts of 30% acid. How much of each must be used?

19. Jones has a 90% solution of boric acid in his pharmacy which he reduces to the required strength by adding distilled water. How much solution and how much water must he use to get 2 quarts of 10% solution?

20. A service station checks Mr. Gittleboro's radiator and finds it contains only 30% antifreeze. If the radiator holds 10 quarts and is full, how much must be drained off and replaced with pure antifreeze in order to bring it up to a required 50% antifreeze?

21. The Dingles have some friends drop in. They wish to serve sherry but do not have enough to serve all the same kind. So they mix some which is 20% alcohol with some which is 14% alcohol and

have 5 quarts which is 16% alcohol. How much of each kind of sherry did they have to mix?

22. A bus leaves Riverside for Springfield and averages only 48 mph. Ten minutes later Smith leaves Riverside for Springfield traveling 64 mph in his car. How long before he overtakes the bus?

23. Two trains leave Chicago, one headed due east at 60 mph and one headed due west at 50 mph. How long before they will be 990 miles apart?

24. A scenic road around Lake Rotorua is 11 miles in length. Tom leaves the hotel on his bike averaging 10 mph and heads west around the lake. Bill leaves at the same time and heads in the opposite direction at 45 mph. How far from the hotel will Tom have traveled when they meet?

25. A small boat sends a distress signal giving its location as 10 miles from shore in choppy seas and says it is making only 5 knots. A Coast Guard boat is dispatched from shore to give aid; it averages 20 knots. How long before it meets the disabled boat if they are traveling toward each other? How far did each travel?

26. The Allisons are on a cross-country trip traveling with the Jensons. One day they get separated, and the Jensons are 20 miles ahead of the Allisons on the same road. If the Jensons average 60 mph and the Allisons travel 70 mph, how long before the Allisons catch up with the Jensons?

27. Three-fifths of the men in a chemistry class have beards and two-thirds of the women have long hair. If there are 120 in the class and 46 are not in the above groups, how many men and how many women are there in the class?

28. The Girl Scouts have a yearly cookie sale. One year they have two varieties which don't sell so well, so they decide to mix them. The coconut macaroons sell for $3 a pound and the maple dates sell for $4 a pound. How many pounds of each do they use so that they will make the same amount of money but have 100 pounds of mixture and sell it at $3.20 per pound?

29. Mr. Higglebotham traveled 60 miles across the rainy English countryside at a constant rate of speed. If it had been sunny, he could have averaged 20 mph more and arrived at Broadmoor in 30 minutes less time. How fast was he driving? (This problem takes a knowledge of quadratics.)

30. A plane flies from Los Angeles west to Paradise Isle and returns. During both flights there is a steady upper air wind from the west

at 80 mph. If the trip west to Paradise Isle took 17 hours and the return trip to Los Angeles took 13 hours, what was the plane's average airspeed?

31. Timmy and Janie sit on opposite ends of a 20-foot teeter-totter. Timmy weighs 60 pounds and Janie weighs 40 pounds. Where would Sally, who weighs 50 pounds, have to sit to balance the teeter-totter? (The fulcrum is at the center.)

32. Jerry needs to pry a 50-pound rock out of his garden. If he uses a 6-foot lever and rests it on a board 1 foot from the rock (for the fulcrum), how much force must he exert to raise the rock? (Assume balance.)

33. Where must the fulcrum be located if a 250-pound weight and a 300-pound weight balance when placed on each end of an 11-foot bar?

34. An 88-pound boy sits on one end of a 15-foot board, 4 feet from the point of balance. His friend comes along and gets on the other side at a position which enables them to balance. How far from the fulcrum will the friend be if he weighs 64 pounds?

35. Tim and Tom sit on opposite ends of a 20-foot seesaw. If Tim weighs 120 pounds and is 8 feet from the fulcrum, how much does Tom weigh if they balance?

36. Mr. Swanson needs to move a 350-pound refrigerator. He has no dolly, so he gets his son to help balance the refrigerator while he slides a 6-foot board under it and uses a block as a fulcrum. (This enables him to get a rug under the refrigerator on which it will slide.) If the fulcrum is 18 inches from the end of the board, how much force is needed to raise the refrigerator?

37. Tickets for the local baseball game were $15 for general admission and $8 for children. There were 20 times as many general admission tickets sold as there were children's tickets. Total receipts were $61,600. How many of each type of ticket were sold?

38. Phineas has $1.15 worth of change in his pocket. He has three more dimes than quarters and two more dimes than nickels. How many of each type of coin has he?

39. Bob has a coin collection made up of pennies and nickels. If he has three times as many pennies as nickels and the total face value of the coins is $16, how many coins of each kind are in the collection?

40. The boys have a small game going in the back room. Mr. X decides to pull out and finds he has $194. If he has four times as many $1 bills as $5 bills, one-fifth as many $2 as $5 bills, and the same number of $50 bills as $2 bills, how many bills of each kind does he have?

41. Jake Zablowski has a collection of coins worth $96.07. He has five times as many 50-cent pieces as he has silver dollars. The number of dimes is twice the number of 50-cent pieces. There are seven more than 3 times as many pennies as dimes. How many of each kind of coin does he have?

42. Zac is 10 years older than Zelda. Twelve years ago he was twice as old. How old is each now?

43. Cassandra is twice as old as Zitka. If 12 is added to Cassandra's age and 6 is subtracted from Zitka's, Cassandra will be six times as old. How old is each now?

44. Maggie and Rod left at 9 A.M. on a hike to Tahquitz peak, traveling at an average rate of 2 mph, with three 5-minute rest stops. After a 45-minute lunch at the peak, they returned home averaging 3 mph and arrived at 3 P.M. How long did it take them to reach the peak and how far did they hike to reach it?

45. Jay's father is twice as old as Jay. In 20 years Jay will be two-thirds as old as his father. How old is each now?

46. Abigail will not reveal her age but says she is 3 years younger than her sister Kate. Ninety years ago Kate was twice as old. How old is each now?

47. Chuck is 22 years older than Jack. When Jack is as old as Chuck is now, he will be three times his present age. How old is each now?

48. The length of a garden is 3 feet more than four times the width. The fence enclosing it is 36 feet long. What are the dimensions?

49. A farmer wishes to fence a circular ring for his ponies. If he has 264 feet of fencing, what would he make the radius of the circle? (Use $\pi = \dfrac{22}{7}$.)

50. The first angle of a triangle is twice the second. The third is 5° greater than the first. Find the angles.

51. Mr. Smith and Mr. Jones each built a similar stone fence to enclose his backyard. They only enclosed three sides, each using his house as the fourth side and had a community fence between the two yards. If the total cost was $5 per linear foot, the yards were 20 feet wider than they were deep, and the total cost was $950, what were the length and width of each yard?

52. The length of a certain rectangle is 4 feet greater than the width. If the length is decreased by 3 feet and the width is increased by 5 feet, the area will increase by 21 square feet. What were the dimensions of the rectangle?

53. John can paint his car in 2 hours, and Joe can paint a similar one in 3 hours. How long would it take them to paint one together?

54. Phoebe and Phyllis run a typing service. Phoebe can type a paper in 50 minutes or the two can type it together in 30 minutes. How long would it take Phyllis to type the same length paper alone?

55. Kel, Del, and Mel are painters. Kel can paint a room in 5 hours, Del in 4 hours, and Mel in 6 hours. One day they all start to work on a room, but after an hour Del and Mel are called to another job and Kel finishes the room. How long will it take him?

56. Bud and Karen went fishing. At the end of the day they compared catches. Together they had caught 40 fish. If Bud had caught two more and Karen had caught two fewer, they would have caught the same number of fish. What was the size of each catch?

57. The Barretts took an automobile trip to Canada. The gas averaged $1.25 per gallon. They spend $60 per night on motels and $40 per day on meals. They average 120 miles per day and got 30 miles per gallon on gas. If the total cost of the trip was $2625, how many days were they gone and how many miles did they travel? (They were gone the same number of days as nights.)

58. A 4-inch-wide picture frame surrounds a picture 2 inches longer than it is wide. If the area of the frame is 208 square inches, what are the dimensions of the picture? (*Hint:* Find the difference between the area of the picture and the area formed by the outside dimensions of the frame.)

59. A rectangular swimming pool is surrounded by a walk. The area of the pool is 323 square feet and the outside dimensions of the walk are 20 feet by 22 feet. How wide is the walk? (This problem takes a knowledge of quadratics.)

60. A lab assistant needed 20 ounces of a 10% solution of sulphuric acid. If he had 20 ounces of a 15% solution, how much must he draw off and carefully replace with distilled water in order to reduce it to a 10% solution?

61. Hank and Hossein open a print shop. They receive a job which Hank can do in 12 hours or Hossein in 14. They start to work on it together, but after 3 hours, Hossein has to stop to finish another job. Hank works alone for an hour, when he is called out on an estimate. Hossein come back and finishes the original job alone. How long will it take him to finish?

62. A reservoir can be filled by an inlet pipe in 24 hours and emptied by an outlet pipe in 28 hours. The foreman starts to fill the reservoir but he forgets to close the outlet pipe. Six hours later he remembers and closes the outlet. How long does it take altogether to fill the reservoir?

63. At the Indianapolis 500, Carter and Daniels were participants. Daniels' motor blew after 240 miles and Carter was out after 270 miles. If Carter's average rate was 20 mph more than Daniels' and their total time was 3 hours, how fast was each averaging?

64. The pilot of the Western Airlines flight from Los Angeles to Honolulu announced en route that the plane was flying at a certain airspeed with a west wind blowing at 30 mph. However, after flying for 3 hours, the wind decreased to 20 mph (airspeed constant). The plane arrived in Honolulu 2 hours later. If the distance from Los Angeles to Honolulu is 2500 miles, what did the pilot announce was the airspeed of the plane?

65. Mr. Monte spent 3 percent of his salary on property taxes, 4 percent went for health insurance, and 10 percent was put in the bank. If what remained of his yearly salary was then $30,500 per year, what was his yearly salary?

ANSWERS

1. 25, 86
2. 12, 36
3. 17, 18, 19
4. 23
5. 81, 83, 85
6. 22, 24, 26
7. 2 feet
8. 27
9. 43
10. 83
11. $28,000, $22,000
12. $30.60
13. $12,000, $14,000, $26,000
14. $900 for first house, $1200 for second, $1200 for third

15. 130 hours for Heinrich, 140 hours for Wolfgang
16. $4800, $4000
17. 2 liters
18. $3\frac{3}{5}$, $2\frac{2}{5}$ quarts
19. $\frac{2}{9}$, $1\frac{7}{9}$ quarts
20. $2\frac{6}{7}$ quarts
21. $1\frac{2}{3}$, $3\frac{1}{3}$ quarts
22. $\frac{1}{2}$ hour
23. 9 hours
24. 2 miles

25. 24 minutes, 2 miles,
    8 miles
26. 2 hours
27. 90 men, 30 women
28. 80 pounds of coconut
    macaroons, 20 pounds of maple
    dates
29. 40 mph
30. 600 mph
31. 4 feet from fulcrum on
    Janie's side
32. 10 pounds
33. 5 feet from 300 pounds,
    6 feet from 250 pounds
34. $5\frac{1}{2}$ feet
35. 80 pounds
36. $116\frac{2}{3}$ pounds
37. 200 children's, 4000
    general admissions
38. 3 nickels, 5 dimes,
    2 quarters
39. 200 nickels, 600 pennies
40. 10, 40, 2, 2
41. 20 dollars, 100 fifty-cent
    pieces, 200 dimes, 607
    pennies
42. 22, 32
43. 12, 24

44. 3 hours, 6 miles
45. 20, 40
46. 93, 96
47. 11, 33
48. 15 by 3 feet
49. 42 feet
50. 35°, 70°, 75°
51. 50 feet, 30 feet
52. 8 by 12 feet
53. $\frac{6}{5}$ hours
54. 1 hour, 15 minutes (or
    75 minutes)
55. $1\frac{11}{12}$ hours
56. 18, 22
57. 25 days, 3000 miles
58. 8 by 10 inches
59. $1\frac{1}{2}$ feet
60. $6\frac{2}{3}$ ounces
61. $6\frac{1}{3}$ hours
62. $23\frac{1}{7}$ hours
63. 160 mph Daniels' rate,
    180 mph Carter's rate
64. 526 mph
65. $36,746.99 yearly salary